MW00582847

SPARKING INNOVATION

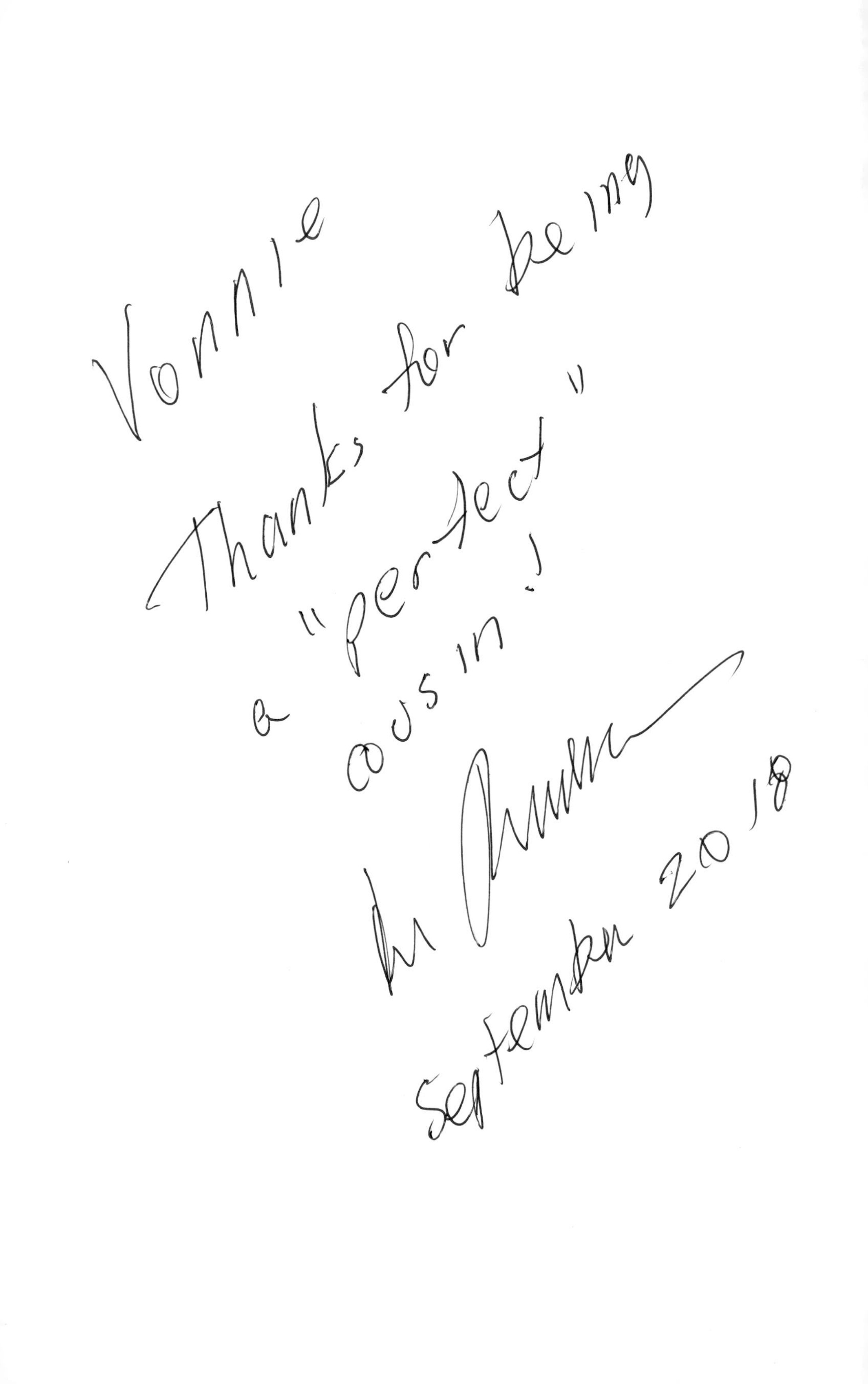

Vonnie

Thanks for being a "perfect" cousin!

[illegible]

September 2018

D. Kenneth Richardson

SPARKING INNOVATION

Lessons to Spur America
to Regain Its Lead
in Science and Engineering

Sea Hill Press

Sea Hill Press Inc.
www.seahillpress.com

First Printing 2018

ISBN 978-1-937720-43-8

Library of Congress Control Number: 2018941828

Printed in South Korea

This book is dedicated to my son

Bruce Kenneth Richardson

A prime exemplar of trust, support, wit, and wisdom

CONTENTS

FOREWORD

My association with Ken Richardson, long and enriching, reflects a much longer relationship of the USC Viterbi School of Engineering and the technology ecosystem in Southern California. Along with our partners at Caltech and UCLA, USC engineering has helped shape, and at the same time benefit from, the fantastic feats in engineering and technology that originated from the vibrant aerospace and defense industries. As president and chief operating officer of Hughes Aircraft Company, and in his long career in the same company, Ken had a front-seat view on the most essential aspects of innovation: on how creativity and discoveries are seeded and nurtured, how they are transformed into maturity with meaningful impact, and how they are applied to useful purposes.

I like to think of technology in its most encompassing basis as "leveraging phenomena for useful purposes"—as brilliantly distilled by Brian Arthur in his book *The Nature of Technology*. "Useful purposes" are infinitely diverse, and they notably capture the breadth of imaginative projects aptly presented in Ken's book. But they also include the discovery of new phenomena, which when entered in the above definition, strongly hints that technology has an inherent autocatalytic effect, the most emphatic demonstration of which is Moore's law (the number of transistors in a dense integrated circuit doubles approximately every two years).

Such analogies suit me perfectly well, given my tendency to view things, including the evolution of technology, with the prism of chemical reactions—after all, I am deeply influenced by my own background! It is this frame of mind that has helped me appreciate more deeply Ken's exciting voyage in discovery and innovation described in the next pages:

from the recruitment and assembly of talent (the "reactants"), to the creation and nourishment of a culture of creativity (the "reactors"), to the enhancement of thought leadership and to the impact of solving old problems and in enabling the solution of new ones (the "products").

The successful practices described stem from Ken's years of experience in the golden era of aerospace. But, they are quite applicable in today's world. Of course, there are fundamental differences between the world now and that during the Cold War. Communications and digital revolutions bringing in exponential changes were impossible to imagine a mere few decades ago. Today, the rate of change is much faster, and tomorrow it will further accelerate. Surely, best practices in innovation will evolve, as they are also inevitably subject to the same rate of transformation. Leading our exponential times (I often use the term "hugging the exponential")—as wonderfully explained by Stanford psychologist Carol Dweck—requires a growth mindset: one that encourages agility, flexibility, entrepreneurial thinking, and constant experimentation. And yet, unsurprisingly, this very same mindset can be traced within the pages of *Sparking Innovation*, even though they were derived at the early onset of our continuing technology evolution. I quote from this book's preface:

> The goal is to inspire innovation and sustain that drive for the long term. The best methods will differ between groups engaged in radically different projects, the type of professionals involved, and the duration of time needed for completing the tasks. Planners must cope with influences posed by continual evolution of our culture, the ambient society where activities will take place, and volatile governmental styles.

This wording, in many ways, portrays a growth mindset.

How relevant is the rich journey recounted by Ken to today's world? In 2002, the National Academy of Engineering (NAE) published a list of the "20 Greatest Engineering Achievements of the 20th Century," from electrification to computers, to space technology, and to high-performance materials. Many of the advances from the defense and aerospace pioneers in this book surely contributed to that list. Then, in 2008, the NAE published a different study, now looking ahead on what are likely to be the Grand Challenges for Engineering of our times. Three of the fourteen challenges on the panel's list are to reverse-engineer the brain, prevent nuclear terror, and secure cyberspace. Most importantly, these challenges are all described by an active phrase, "engineer the tools of scientific discovery." This call to action will be accomplished by the empowering nature of engineering to progressively enable other disciplines and the sciences, what I like to call Engineering Plus. It will require no less in innovation, ingenuity, talent, and the nurturing of ecosystems than described in this book. But with one important difference: The engineers who will solve these challenges will reflect a changing of the conversation about what we do in engineering, who we are, and what we look like. It will be the enthusiastic and vibrant class of the engineering students of our times—diverse, innovative, and brilliant. One can't wait for this next chapter.

Yannis Yortsos, PhD
Dean Viterbi School of Engineering
University of Southern California

PREFACE

FOR THE LAST one hundred years, this nation rated first in achieving advancements in virtually every segment of technology. During the current century, that leadership position is slipping. The 2002 United Nations overall rankings on innovation achievements placed us in second place to Finland. According to the Global Innovation Index, in each year from 2014 through 2017, the U.S. has ranked no higher than the 4th most innovative country in the world. What is particularly striking is the U.S. is listed as 41st in education quality and content, as well as an appalling 85th in percentage of graduates in science and engineering. We must strive to regain our competitive position in the world. To do this will require the active participation of individuals and teamwork efforts, uniting the government, industry, and the public. The lessons of historic successes described in this book may provide useful guidelines for the positive steps to be taken. Underlying all is an attitudinal change in our culture by placing a much higher priority on supporting technology endeavors. This can be accomplished by increasing funding investments, improving educational content, and motivating a higher proportion of our youngsters to pursue professions in technology advancement.

A technical career offers unlimited opportunities for a person trained in science or engineering: to discover, to create, and to change the world for the better. You can be well rewarded for doing work that doesn't feel like work at all! Many talented people already active in a career may be a bit unsure on how to focus their abilities to satisfy current and future needs of society, businesses, and governments. The contents of this book provide useful clues for realigning skills or adding training so as to begin a better path toward achieving personal and career ambitions. These suggestions are combined with brief overviews of several historic successes to help the reader select a winning long-term balanced education that will provide a rewarding and satisfying career. The thoughts herein are based on personal observations made during active participation in forty years of the golden era of aerospace as well as from a lifelong interest in history. Descriptions of personal experiences are intended to authenticate the reality of the operational recommendations being stated rather than as autobiographical memoirs. Budding entrepreneurs will also be able to better perceive the many activities and problems that must be conquered in establishing a new business enterprise. As Thomas Carlyle said in 1840, "The present is the living sum-total of the whole past." This phrase certainly applies to technology advancement. Almost all new things rely on a solid foundation of accumulated knowledge. It is hoped that the snippets of history captured in this book will serve to help and inspire both young and seasoned scientists, engineers, and leaders to get the most out of their careers.

The underlying theme of this text is finding ways to stimulate, nurture, and inspire creativity. Observing inventions of the past as well as recent positive management styles can render useful insights into how best to move forward. The *Encyclopedia Britannica* defines creativity as "the ability to make or otherwise bring into existence something new, whether a new solution to a problem, a new method or device, or a new artistic object or form." Its description of innovation states: "the creation of a new way of doing something, whether the enterprise is concrete (e.g., the development of a new product) or abstract (e.g., the development of a new philosophy or theoretical approach to a problem).

The *National Geographic* magazine in June 2017 opined that the world's top ten innovations of all time are: (1) printing press, (2) light bulb, (3) airplane, (4) personal computer, (5) vaccines, (6) automobile, (7) clock, (8) telephone, (9) refrigerator, and (10) camera. This author feels this is a worthy judgment, but the list could equally include: agriculture, the wheel, monetary systems, religious and ethical credos, sea-worthy vessels, the telescope, railroad networks, electrical power, radio, satellites, harvesting nuclear energy, and the global Internet. Even a ranking of the top one hundred is an enormous challenge, and selection is strongly influenced by personal experiences and attitudes. Many vignettes about marvels of creativity in hardware during the old days will be found in several of the following chapters. Let us hope that a future *National Geographic* issue will compose another honored grouping of giant jumps in verified conceptual theories. Surely it will include past discoveries in gravitational forces, electromagnetism, thermodynamics, evolution in nature, relativity in the universe, as well as the detailed structures of atoms and organic genes.

Scientific training and expertise are the expected sources for deriving and proving how things exist or function, using understanding of physics, chemistry, biology, and other measurable aspects of nature. Hypotheses and theories to diagnose and solve knotty problems can usually be precisely expressed using the tools of mathematics. Verification of these concepts is done using physical experiments conducted in well-controlled laboratory environments. The engineering professionals then use practical techniques to convert these theories into useful objects and methods. However, history shows that innovations have sprung from clever people in all fields of endeavor, not just from technical wizards. Many new things are hatched from surprise events; some as remedies to overcome urgent needs, others from an innate desire to create something never previously done. Herein we will suggest some guidelines for coaching and inspiring teams of individuals in a business to enthusiastically set meaningful goals as their devoted way of life. It is especially hoped that some of these suggestions may spur U.S. corporations to accelerate their progress in our very competitive world.

Finding optimal ways to establish and manage large organizations doing complex work can be very difficult. The primary objective should be to inspire innovation and sustain that drive for the long term. The best methods will vary between groups engaged in radically different projects, the type of professionals involved, and the duration of time needed for completing the tasks. Planners must cope with influences posed by continual evolution of our culture, the ambient society where activities will take place, and volatile governmental styles. Corporate flexibility must emphasize a long-range philosophy and a properly funded research and development effort. A strong foundation can be learned from understanding what has worked well in the past from five noteworthy books recommended in chapter 3. Further ideas about winning practices are described in several other chapters. Six books cited in chapter 4 provide further visions of today's successful practices for stimulating innovation. Collectively this text and those references outline suggestions for perpetuating a positive operating environment, embedding an inspirational philosophy, properly clustering diverse skills, strongly welding closely-knit teams, and correctly selecting good coaches. Paramount at all times is to maintain and assure employee satisfaction in agreeable task assignments, achievement recognition, reward for performance, and continual potential for career growth. A primary goal must be molding an atmosphere that encourages independent thinking and open communication.

Included herein are many bits of history showing how entrepreneurs and business leaders successfully practiced these arts. Numerous examples of how devices commonly used today came about are described, some emanating from surprising sources. Comprehending these innovations yields key hints about how to properly coach any activity. With a solid knowledge of historic successes, imagination thrives.

Very special and dynamic operating challenges must be tackled by managers exploiting the mystique of the complex electromagnetic spectrum, as we at Hughes Aircraft Company did so well. Fortunately, almost any firm comprised of engineering, manufacturing, customer user support, marketing, financial, and legal functions can emulate those practices. Smart adaptation to the particular skills needed in their enterprise will payoff well. Everyone must cope with surprises that always seem to pop up as we strive to push beyond conventionally perceived boundaries. Several chapters will suggest ways to deal with high stress, volatile changes in competition, antiquated government restraints, and the rapid changes in understanding the depths of technology. In our company, fulfilling vital national goals required much attention and dedicated vigilance. You will find in chapter 12 the author's four most memorable career assignments in contributing a role in fulfilling our country's objective. Dramatic proofs-in-the-pudding attest to the significance of our firm: the U.S. prevailing in the forty-five-year-long Cold War in which we countered the Soviets without military combat; and the eighty-eight Hughes systems that were key elements in assuring a quick triumph in the 1991 Gulf War against Iraq.

Neanderthal

Active Brainpower

Space Walk

We all should keep in mind the reasons this nation was so successful during the golden era of aerospace. Emulating some of the overall rationale, perhaps, could place us on a better path in our rapidly changing competitive world. Here's why we all did so well: The government established two significant national missions that unified the citizenry and supported large federal investments. Those were "prevail in the Cold War" and "win the space race." Then, Washington set up many technological programs designed to achieve those goals. Adequate funding was provided, and key projects were competed among many competent domestic design and development corporations. Industry responded and supplemented the government funding with assets from other sources to further increase the investment in R&D. (This also improved their competitive posture.) The results were astonishing and placed Hughes first in the world in most fields of technology innovation. Those national missions engendered a robust economy, attracted new talent into technical professions, and created many lifestyle improvements (examples are the Internet, cell phones, GPS, and worldwide communication.)

Humankind has come a long way in advancing technology and will continue to invent lifestyle improvements by providing an inspiring environment for our brainpower. Perhaps this book will provide some foresight and guidance to future innovators and leaders in the United States.

1

CLASSIC INVENTIONS

WE TEND TO think of invention as something new, something that will change our future, hopefully for the better. Yet the mental processes that lead to new ideas are as old as mankind. Many of the amazing inventions of the past have become such a part of our lives that we take them for granted. However, in their time they were unexpected, fundamental, and even radical in nature.

There's a lot to be gained by surveying the classic inventions of history. We can see how they span an amazing range of human needs and skills, and we can marvel at the elegant way they solve previously intractable problems. Best of all, by imagining the mental processes leading to these brilliant solutions we may be inspired toward new approaches and new ideas to tackle problems facing us now and in the future. This chapter describes some remarkable past inventions, both large and small, that the author believes can stimulate creative thinking today.

The Many Faces of Creativity

As mentioned in the preface, *creativity* is the ability to make or otherwise bring into existence something new, whether a new solution to a problem, a new method or device, or a new artistic object or form. Brilliant ideas can come from any source and can be a sudden inspiration or the result of dedicated efforts by an organized staff.

Here are two examples of breakthrough ideas coming from unexpected sources. In the early twentieth century, thousands of workers were imported to Hawaii from Japan and the Philippines to maintain and harvest the sugarcane plantations. One sugar mill product was burlap sacking that was made from cane fibers called "bagasse." These bags were used to store many farm products and were most abundant. An Okinawan named Zenpan Arakawa noted that clothing for many workers was quite scarce: why not make shirts from surplus or used burlap? This new bagasse concept was very successful, resulting in the emergence of the iconic century-old "Arakawa Store" (unfortunately, it closed in 1995) and the shirt morphing into a now-famous symbol of Hawaii: the denim **palaka shirt** with its plaid decoration. *Palaka*, Hawaiian word for plaid, fabrics originated in the early 1800s, while King Kamehameha I ruled. But Arakawa's hard work boosted them to universal public demand in those islands.

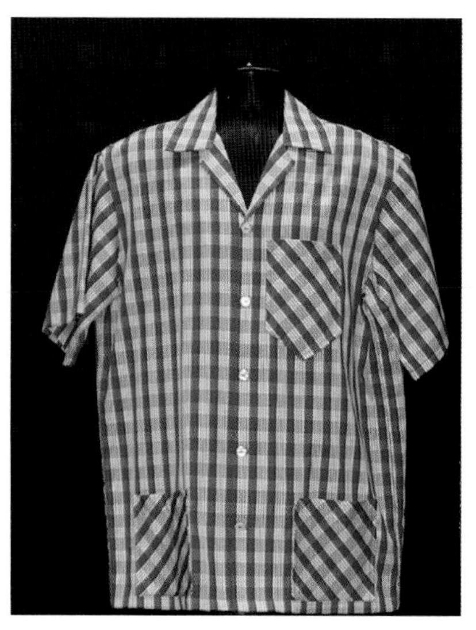

Traditional *Palaka* Shirt
(courtesy of Alohaland.com)

Another unexpected invention was the birth of the glass elevator, an architectural attraction for many tourists in cities with fascinating views. When elevators became competitively necessary to ease the comfort of hotel clients, existing buildings had many internal structures that limited vertical shaft spaces from being added. In San Diego, a team of experienced architects and engineers conferred for two weeks and were unable to solve the installation problem for the El Cortez Hotel. The building's porter observing this dilemma suggested, "Why not just mount it on the outer wall." This outside-the-box thinking, unconstrained by conventional professional training and procedural boundaries, won the day. It certainly embarrassed the experts, and the exterior **glass elevator** soon became a very desirable method for hotel elevator placement.

Glass Elevator at El Cortez Hotel
(courtesy of San Diego History Center)

In contrast to this illustration of constrained thinking by highly educated professionals, an advanced education environment is often an incubator for conceiving something new and worthwhile. In 2016, a group of engineering students at the University of Southern California competing in an annual entrepreneurial contest came up with **Nose Knows**, a method of individually identifying dogs. Patterns in each nose are unique, similar to the fingerprints of humans. Nose Knows photographs a dog's nose and places those digital images in central storage. A lost dog has its nose photographed; the image is inserted into the computer center, which matches it to the existing data base. Eureka! My dog has been found!

Doggy's Distinctive ID

The thought processes that spurred each of these examples are different, but equally illustrative of ways to achieve a mindset toward creativity: Arakawa saw a critical need for clothing that could be remedied by using surplus materials; the porter became aware of a desirable objective that could be attained when not being constrained by procedural habits; the USC team searched for a new way to apply advanced technology they had just learned. Many new devices can be a different application of previous inventions or an aggregation of successful functioning elements.

Easing Heavy Burdens

The device we call the **wheel** has allowed an astonishing number of improvements in human lifestyle, making our physical efforts far easier. One can make a long list of such things: motion of heavy objects over hard surfaces, generation of power from water or gas streams, conversion of rotational rates or transfer of energy from one device to another, and easing the reshaping of malleable objects in a circular manner. A wheel is constrained by an axle shaft. The center of some disks is firmly fixed to the shaft. That shaft passes through one or more bearings that are affixed to the support frame, allowing rotation. Others have a central hub containing a bearing, and the shaft is rigidly attached to the frame.

Artwork and written records indicate that the first useful rotating wheels were devised about 4000 BCE in the Levant. These were disks of wood, spun horizontally, and were used for creating artful pottery. Wheels using clay and stone soon followed. When appropriate technology became possible, they were fabricated with several types of metal, plastic, or fiberglass. Within the next few centuries, the idea occurred to have the disks rotate in vertical directions. This then allows load-bearing vehicles to aid in agriculture and transportation. This new application seems to have sprung forth about the same time in Mesopotamia, central Europe, and the Caucasus. In about 2000 BCE, spoked wheels were invented. These were lighter weight, stronger, and more varied in the material types used. Egyptian and Syrian frescos dramatically illustrate **spoked-wheel chariots**, giving them an almost invincible agile vehicle for successful military encounters.

Historical records do not seem to credit any one person for conceiving and implementing that first spinning potter's wheel, nor an individual who first turned the axle horizontally to carry useful loads with ease. It also seems credible that the latter inspiration may have cropped up at about the same time within several different cultures that were remote in location and had no interactive communication. This author is also astounded and perplexed that Orville and Wilbur Wright did not use wheels instead of skids as the undercarriage of their record-setting *Flyer*. They were already accomplished in designing, manufacturing, and selling bicycles. The obvious application of wheels occurred to their archrival, Glenn Curtis. Competition and patent rights by Curtis further impeded the Wrights from adopting this perfect method for aircraft weight-bearing with mobility on the ground.

Syrian War Chariot

First Flying Machine

The **boomerang**, iconic symbol of Australia, was first observed in action there by Europeans in 1804. A 50,000-year-old rock inscription by Aborigine tribes depicted those amazing devices. Much evidence shows boomerangs evolved independently in, or migrated to, most parts of the world, including King Tutankhamun's fourteenth-century BCE collection. The oldest physical remnant was found in Poland dating back to the Stone Age, 30,000 years ago. Usually made of wood, and being up to six feet long, they were used for hunting small game and birds, sport competitions, musical performances, and as toy playthings. What seems astonishing today is that these are aerodynamic wonders created by people with no knowledge of what we now call physics, mathematics, and hardware engineering. The first versions were throwing sticks that could be propelled with great speed in a spinning motion, flew in a straight line, and struck with great force. Both arms were symmetrical, with carefully shaped airfoils that provided lift with little drag as the device spun through the air. These sticks were also substantial enough to be used as sharp hand clubs in close personal combat. Historians speculate that

Australian Boomerangs

as individuals tried to hone the wood shapes to assure straight flight, they discovered nonsymmetrical arms could alter the flight path. Careful shaping can cause the differing lift of each arm and the rotational spin axis to be other than vertical. This will cause the returning boomerang to travel in an elliptical path back to its departure point. This beauty is a fantastic Homo sapiens' creation of the first man-made heavier-than-air flying machine.

Finding Our Way

Artifacts of a **magnetic compass** have been found to date over two thousand years ago in China. Uses were for accurate placement of new building construction and by soothsayers as a tool to help predict fortunes of individuals or about chancy decisions being contemplated by uncertain groups. The material was naturally occurring lodestone, composed of magnetite, an iron oxide aggregation that we now believe to have been polarized by the strong magnetic force surrounding lightning bolts. When properly suspended to swing freely around its center, it will rotate to align with Earth's magnetic orientation.

During the tenth century CE, the Chinese created the first compass to be used for navigation. Some written records indicate that one configuration was a magnetized iron needle floating in a bowl of water. Another design, used during the next seven hundred years had a lodestone hanging in an upside-down wooden bowl that could rotate so as to point north. That device was first used by army land forces to orient themselves at night. The first appearance of a position-indicating dial beneath the iron needle *(compass rose)* in seaborne compasses was described in thirteenth-century Chinese historical documents. At sea, it has always been difficult to sail vessels using directional courses beyond sight-of-land features after dark and when foul weather obscures long-range vision. In better environmental conditions, steering could be done by sun and star sightings. The new compass provided an angle of reference (north) to keep steady on the boat's desired cruising line. Navigational use appeared around the twelfth century CE in Europe. The technology was probably transferred from China by way of the Arab traders. Thus began the many design advances since then to make today's **dry compasses** lightweight, accurate, and of low cost. Extremely precise floating "wet compasses" are gimbal mounted for boat motion compensation aboard many seaborne vessels. Historical documents do not cite specific individuals for inventing any breakthroughs in implementing this most useful device. An observer today can savor that two thousand years of active thinking and experimenting finally overcame a significant hazard to ocean navigation. International trade growth and personal survival motivated many people to compile good results by their predecessors into a composite success.

Pocket Compass

Some people were confused about apparent mismatches of navigation charts to what the compass was measuring. Not understood until about 1700 CE was that magnetic north differs from the Earth's true geographic north. Those angular deviations differ for each spot on the Earth's surface. During the nineteenth century, accurate numerical offset values for many known sites were finally published in nautical charts and almanacs for use by mariners. The deviation during 2017 in Los Angeles was 13.6 degrees east. Such declination changes annually since magnetic north results from shifting electrical energy produced within the Earth's spinning liquid core. It is obvious that adjusting for the local declination is vital for precise long-range travel.

Not Missing a Beat

During all of mankind's existence, there has been a pressing need to measure time. Survival often depended upon forecasting the seasons for crop growth and animal migration, foretelling the phase of the moon, predicting tidal cycles, benefiting from daylight duration, sunlight strength expectations, coordinating individual and group actions, and tracking the ages of humans and livestock.

Thousands of years ago the Egyptians became adept at defining time: one year elapses between the cyclical northernmost angle of the sun; there are approximately twelve full moon phases each year; and each day consists of twelve daylight hours plus twelve hours of night. Each of these hours varied with the continual change with the Earth's rotational axis tilt; the twenty-four increments were only identical at each latitude on the spring and fall equinoxes. The choice of twelve is speculated to be either because of the annual moon cycle or that culture's practice of sequentially counting by tapping three joints of four fingers on a hand using a thumb. Greek philosophers decided that, for mathematical precision, each hour must always be equal in duration rather than varying with Earth's continually changing axis tilt. The Babylonians divided the hours into sixty minutes, each composed of sixty seconds. The Sumerians in 3500 BCE had decided to do mathematical counting of large amounts by sixties, instead of our hundreds. (Perhaps it was because sixty results from five counts of twelve.)

Most ancient methods were to observe sun angles by noting shadows cast by fixed objects and by measuring star positions at night. Vertical stones and wooden rods formed sharp sun shadows that were longest at dawn, shortened at a regular pace until noon, then lengthened moving eastward until sunset. A noontime shadow with no east-west image will point north or south, giving a measure of the season. Relics of such shadow measurers can be found in England's Stonehenge as well as obelisks in Egypt, Iran, Iraq, and China. Today we find a familiar time machine is the **sundial** (oldest evidence found was from 1500 BCE Egypt). Atop a circular disk is a triangular arm with a straight edge. The disk is mounted horizontally and the vertical arm is aligned to Polaris, the North Star. As the sun makes its trek, the shadow of that straight line falls on stripes on the marked dial beneath, denoting the hour. Similarly, an **armillary** uses a slanted rod above a circular ring with markings denoting hours when

Armillary Timepiece

the rod's sun-shadow crosses them. Another clever idea was to allow sand streams to fall from top to bottom through a narrow opening in what we all know as an **hourglass**, shaped like a figure eight (once again devised by those smart 1500 BCE Egyptians). Its reset merely requires the vessel to be inverted to restart the sand stream. These could be quite accurate, and the container could be large enough for measuring relatively long time spans. Hourglasses have a significant advantage over sundials, being able to operate both night and day. Other gadgets commonly used throughout the past centuries were igniting candles, incense, wicks of oil, or time sticks, each with known burning rates.

Also about 1500 BCE, **water clocks** seem to have existed in Egypt (invention claimed by Amenemhet), the Middle East, and China. These sensed minutes by observing the rate water passed through calibrated holes. One interesting technique was to float a bowl with few holes in its bottom on water in a larger container. Time was measured by how long it took to sink. The bowl also could be marked on its side to divide the total time span into smaller increments. Resetting was easy (you even got to wash your fingers), and little water was lost.

More finite measurement of short time-bits finally became possible with the appearance of an **escapement wheel**. This device converts a continuous power source into a sequence of exactly timed stops and starts. The initial method let a small stream of water drip into the ladle of a pivoting spoon. The spoon's handle has a small protruding latch key. When sufficient weight has accumulated in the ladle, the handle rotates and lifts that latch from a slot in the circumference of a disk, permitting that wheel to turn. The disk's axle shaft is powered by another water stream. Spoon pivoting causes the water to fall from the ladle and the latch to be inserted into another rotation-stopping notch. This oscillating cycle is repeated at a consistent rate and the slots in the disk's edge are equally spaced, thereby measured time intervals. Early versions of this concept are hinted at in Greek writings of 250 BCE and well recorded in eighth century CE China. Tower clocks using similar designs became prevalent throughout Europe during the Middle Ages.

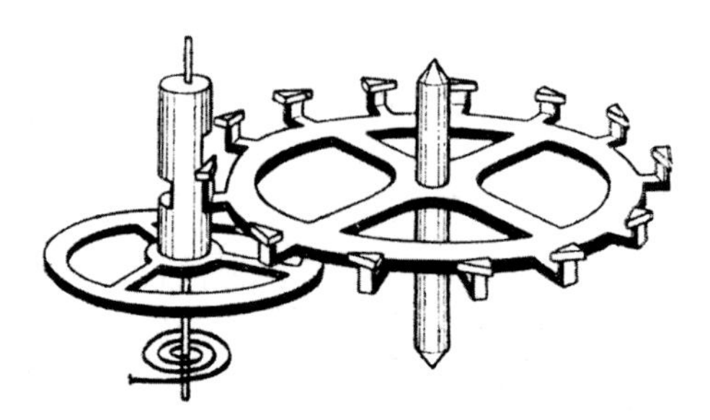

Antique Brass Clock Mechanism

The real technology breakthrough came in Europe in 1656 with the use of a pendulum as the oscillator moving the latch key. The regularity of this "heartbeat" was determined by the physics of pendulum swings: the rate is determined by its length. Power from a manually wound drive-spring or mainspring, created in the fifteenth century CE, forced escapement disk rotation in one direction. An arm at right angles atop the pendulum's pivot has a protruding latch shape at each end.

As the swing reaches its limit, the latch on the opposite end of the arm is inserted into one notch in the wheel (now called a "balance wheel"), causing it to stop. When the pendulum returns in the opposite direction, that latch key is lifted, the wheel turns by a small angle until stopped by the key on the other end of the arm. Linked to the axle are sets of gears to convert the rotation proportionally to three needles depicting seconds, minutes, or hours upon a marked clock dial.

A **balance spring**, conceived by England's Robert Hooke and perfected by Netherlander Christian Huygens in 1657, replaced the pendulum as a time-consistent oscillator for portable clocks and watches. This flat spiral coiled and unwound, repeatedly inserting and lifting a latch key at its outer end into the escapement wheel's notches. When first installed, the spring is slightly compressed when the latch is inserted. With the mainspring wound, the wheel tries to turn. That force further compresses the spiral, causing an outward thrust to lift the key from the groove. The resulting release of the turning pressure causes the balance spring to return to its installation coil shape, thus implanting the key into the next slot. That cycle then repeats. Oscillation rates can be adjusted by moving a "regulator" clamp that changes the spring's active length.

Time fixing is critical to well-executed sea navigation. For many years, star angle observation with astrolabes and sextants yielded reasonable measures of latitude. But knowing one's longitude was far more difficult. Galileo tried to codify observations of Jupiter's four moons whose relative positions varied when viewed from different longitudes. This technique was most impractical for common use. It is notable that most early voyages were done using optimum wind conditions, but for safety's sake, also were planned to steer along a single latitude line until some part of the other continent was found by visual sighting.

H-4 Sea Watch

In the eighteenth century, more precise longitudes could be calculated using the **H-4 sea watch** perfected in 1769 by Britain's John Harrison. He created the first instrument to be almost immune to the environmental strains and continual motion at sea, as well as being stable while the drive-spring was being rewound. Noon can be fixed each day at sea by sextant angular observation or shadows cast by vertical poles. For many weeks, H-4 accurately kept precise track of Greenwich Mean Time (GMT). Upon sextant verification of local noon, a direct comparison with GMT noon gave an accurate longitude fix. The time spread showed the differing portion on the Earth's twenty-four-hour rotation from the known longitude at Greenwich, England. This also can yield a distance traveled: during each four-minute time interval, the equator spins one degree of arc, or sixty nautical miles. One's distance can be derived by multiplying time sailed with the ratio of the current latitude circumference with that at the equator.

With the maturing of electronics technologies, many more ways were found for extremely accurate time measurement: quartz crystal oscillators, digital watches (see chapter 11), computer sequencer regulators, and the atomic clock.

Melding Art and Science

In considering ways to stimulate technical innovation, an attitudinal motivator can arise from observing that history's many successful inventions were a melding of science and art. Perhaps this is because what appears to an appreciative eye is that which is harmonious with nature. This is particularly true with efficient aero- and hydro-dynamic shapes. One sees pleasing contours, smooth interstices between connecting segments, and polished surfaces. These features are compatible with the laws of physics as they apply to motions in air and water. An example is the ancient boomerang described earlier, but here are two more recent shapely sights: the speedy American **clipper ship** of the nineteenth century, and the superb British **Spitfire** fighter.

American Clipper Ship *Challenge*, Built 1851

Those handsome ships were a derivation of the Baltimore Clipper created in the War of 1812. The Cutty Sark, its final version, was built in Scotland in 1869. High speed was made possible with a slim and slick low-draft hull with a small cargo capacity and a plethora of artistic sails that could be rapidly maneuvered to maximize capture of varying winds. The desire for best speed was motivated by contests in transporting tea from China to England. The annual winner garnered higher prices for its cargo than that of its competitors. The *Spitfire* fighters were superior to all others during the Battle of Britain, with altitudes reaching 34,000 feet and speeds over 360 miles per hour. It featured an astounding climb rate,

Supermarine *Spitfire*

and unbeatable high-g maneuverability. Over 20,000 were produced. The author opines that the Supermarine *Spitfire*, designed in the mid-1930s by England's R. J. Mitchell, is the most dashing and attractive aircraft ever built.

Another device whose technical design is harmonious with its operating environment, as well as being an artistic display, is the **airship**. These can ascend from the ground because their interior envelope contains a gaseous substance that is lighter than the ambient air. During the first half of the twentieth century, large airships called "dirigibles," with a passenger capacity nearing one hundred, carried travelers between continents. These had interior skeletal structures that enabled propulsion engines and steering controls. Some nations employed them for anti-submarine warfare in both world wars. The most famous commercial dirigible was the German *Hindenburg*, manufactured by Graf Zeppelin; it burned and crashed in New Jersey after a two-and-a-half-day flight from Berlin in 1937. This vehicle used extremely flammable hydrogen as its lifting element. Prior to that disaster, airships had safely transported 34,000 travelers between the continents. Non-dirigible balloons also using hydrogen were frequently used in the U.S. Civil War and in World War I as observation platforms. These were tethered to the ground and, when threatened by hostile attack, were hastily cranked down by a winch to escape destruction. Since then, such craft use low-density helium, which is chemically inert and therefore not flammable. Hot-air balloons, carrying as many as nine passengers, are very popular today for tourist journeys and competitive air races. A propane burner heats ambient air, thus reducing its density, which is funneled into the lifting envelope.

Periodically, as it cools while aloft, the burner is relit to assure continued flight. These vehicles are subject to the vagaries of wind currents for course direction, and altitude is controlled by those reheats and dropping ballast weights.

Montgolfier's Hot Air Balloon

The first documents describing lighter-than-air craft date from third century CE China. Zhuge Liang used them for military reconnaissance. The first manned flight was in 1709, achieved by Father Bartolomeu Lourenço de Gusmão in Portugal. This no-cargo bird captured hot air in an enclosure made of paper materials. Giant strides in perfecting balloon designs were done in France during the eighteenth century CE. Joseph-Michel and Jacques-Étienne Montgolfier's hot-air balloon demonstrated for King Louis XVI of France the world's first lofting of passengers (a sheep, a duck, and a rooster!). A tethered Montgolfier balloon had already demonstrated in 1783 the world's first lofting of humans; three men were aboard for a public appearance in Paris in August 1793. That same year, Jacques Charles, with Nicolas-Louis and Anne-Jean Robert, created the first hydrogen-filled balloon. It was made with silk fabric, sewn in a spheroidal shape. The gas bag was sealed with rubber that had been softened by turpentine into a paste. The hydrogen was derived from pouring 500 pounds of sulfuric acid onto 1,000 pounds of iron. That

December, this miraculous free-flying bird successfully carried two passengers twenty-seven miles from Paris. All these lighter-than-air fliers were fine examples of technology advances done in artistic manner. The ingenious inventors in eighteenth-century France were likely motivated by competition with their peers and to gain lifetime prestige.

A most artistic and colorful symbol of the Old West is the **tepee**. This simple and clever design was used for centuries by many indigenous inhabitants in North America's Great Plains. Similar structures have been found in northern Europe and Asia. Common to these groups was a nomadic existence, with their primary food source being animal hunting. Harsh conditions necessitated creating shelter that could withstand hostile climate changes. However, these cultures had few tools to make life easier. "Tepee" is a derivative of Lakota Sioux *tipi*, meaning shelter.

Tepees, Symbol of the Old West

The strong structure uses a dozen or more thin poles made from lodge pole pine or red cedar. These were bound together a foot from one end by rawhide strips. When erected, the longer ends were twisted to intercross the top ends and were then spread to form a ten-foot circle. The twisting secured the bindings of the roof joint. Upon this cone was placed a shaped-to-fit covering made with rawhide-sown bison hide; its base circle was fastened to the ground using wooden stakes. The hide's inherent insulating quality proved effective as shelter from strong sunlight as well as icy conditions. An entry opening hole had a flap to close at night. The finished shape was perfect to minimize wind forces and shed rainfall. Two exterior poles were fastened at their tops to a large flap that partially closed the conical opening at the tepee's top. They could be repositioned for wind direction or completely shut for rainfall protection. The roof aperture was necessary to allow smoke to escape from a central fire pit. When collapsed, several shelters could be bound together and transported as a travois pulled by a dog or a horse. Here again, the physical design had engineering excellence and simplicity, all resulting in something pleasing to the eye and a perpetual heritage signature for that culture.

Umbrellas and **parasols** provide protection from sunrays and rainfall. They are light-weight structures of flexible rods fanning out from a fixed or telescoping pole and supporting a woven fabric surface. The canopy can be collapsed, and those for carrying by hand can be small enough to fit in a lady's purse. Many parasols, used for shelter from the sun, are often lavishly colored as a mark of fashion or prestige. Umbrellas for rain protection are usually more understated and less intended to capture attention. These devices are commonly artistic in shape and proportion, an impressive blend of appearance and satisfying the engineering demands: mechanical strength, lightweight, and ease of operating from a full open spread to a miniature portable bundle. Such devices may also be called sunshade, rainshade, gamp, brolly, or bumbershoot.

Asian Parasol

The oldest written records and an artifact found date to first century CE China. Some bronze castings of mechanisms similar to parasol hinges were found in China that date back to the sixth century BCE, giving a hint of an earlier creative origin. After 100 CE, the collapsible configuration idea soon reached to Korea, then Japan. Today's bamboo-rib parasols in Nippon are virtually identical to those described two thousand years ago. They are often viewed as a symbol of their culture. If you visit Japan, you will immediately see that tour groups are led by a guide displaying an umbrella with a distinctive identifying color. If more are distributed to others in that group, those will be of an identical color so nobody can mistakenly mix with another group. Sculptures found in Persia and Egypt indicate parasols may have arisen even earlier than those written texts and artifact finds in China. A most utilitarian invention from antiquity, umbrellas and parasols can now be found anywhere on the globe.

Sydney Opera House and Harbour Bridge

Art and science also blend well in construction of buildings, memorials, and useful hardware devices. In this case, engineering methods to assure physical strength and long-term stability result in shapes that reward the eye. Two examples are the iconic **Sydney Opera House** and the **St. Louis Gateway Arch**. Australia's objectives for its formal entertainment center, the Sydney Opera House, designed by Danish architect Jørn Utzon, were to point to the world that their country artistically matched all others, as well as to shape the structure to achieve acoustic perfection. Although built well in 1959, experts in 2016 judged that significant acoustic improvements could be done with modern equipment, requiring a long duration shutdown for interior rework, leaving the acclaimed exterior unchanged.

Fashioned by Finnish-American architect Eero Saarinen, the spectacular Gateway Arch in Missouri symbolically commemorates the 300,000 pioneers who in the mid–nineteenth century began a hazardous three-month wagon journey to Oregon or California, seeking a better life. Built in 1963, the structure is physically ideal to withstand abusive winds and adverse weather, and at the same time symbolize that historic long trek. Inside the flowing shape is a sharply curving passageway for small capsules to transport tourists to the viewing chambers at the top of the span 630 feet above the ground. Participants in that exciting ride are packed closely together and can imagine they just left Earth in an outer-space machine.

Gateway Arch

Golden Gate Bridge

Designated by the American Society of Civil Engineers as one of the Seven Wonders of the Modern World is the San Francisco **Golden Gate Bridge**. After more than four years of construction, it opened in 1937. Until 1964, it boasted the longest suspension span in the world, with that center section reaching over 4,200 feet of ocean water. In 2017, it became number fourteen, outranked by bridges mostly in Asia. The leader is Japan's Akashi Kaikyō Bridge, with an awesome span of 6,532 feet. It opened in 1998, but it will soon be superseded in rank by Turkey's Çanakkale 1915 Bridge, which will span 6,637 feet. Suspension cabling construction is the most effective means of safely supporting elevated vehicle roadways and rails over a long distance. Two trestle-strengthened towers are securely planted on the sea floor. Attached to their tops is a very strong composite cable, forming a natural catenary shape. From it hang many small vertical cables that support the roadway. This structure is most efficient in the application of physics and engineering, handling large high stresses caused by adverse weather. It is obvious to any observer that the Golden Gate is a true art form and is an historic signature symbol for San Francisco.

The Roman Empire provided many wonders in engineering, architecture, construction methodology, structural materials, durable roadways, advanced agriculture, sculptural and mosaic artistry, public newspapers, the first bound books, and the modern calendar. Viewed two thousand years later using current cultural standards, the Romans are judged to be too aggressive, tolerant of violence, exploitive of slavery, overly rigorous in social practices, and are faulted for worshipping tyrants as rulers. However, many of those wonders just cited likely germinated in

Pont du Gard Aqueduct Span

the minds of that large polyglot amalgam of talents, whether citizens, foreigners, captives, or slaves. Observing this rich history gives two guidelines on how to nurture creativity today: (1) use previous technical advances as a building block to achieve new goals, and (2) engage motivated team members to tackle complexities together. Most remarkable are the aqueduct systems used to supply water to towns and agriculture. These marvels employed astonishing surveyor and measuring skills, carefully shaped stone blocks, new formulas for concrete, and construction with strong semicircular arches. (This design is now called the Roman arch and is used worldwide.) These were so well made that many are still in use today, or are symbolic artifacts protected by nations that covet them.

One that should be on everyone's travel list is the 19 BCE **Pont du Gard** in southern France. At 160-feet high and currently 902-feet long, it spans the Gardon River, forming part of a 31-mile-long connection from water springs to the bath centers, fountains, and homes of Nîmes. Gravitational flow supplied over one million gallons each day. It is astonishing how accurate the entire system is. In that lengthy 31-mile link, a large portion underground, the descent was only 41 feet. There is only a 1-inch sluice drop across the artistic and precise bridge.

Leonardo da Vinci is regarded today as the most ingenious and creative individual of all time. He displayed conceptual or instinctive knowledge in virtually every field of art and science, and he is often rated as the finest painter and sculptor in history. His scientific inventions were far ahead of their time: many inventions, well thought-out and diagrammed, could not be physically implemented until new fabrication methods,

Leonardo da Vinci's Fixed Wing Flyer

such as metallurgy skills, became practical. A few of these are the helicopter, fixed-wing aircraft, armored vehicles, an adding machine, automated weaving, and work-saving agricultural machinery. Each of these was technically correct in engineering, but also was aesthetically pleasing to the eye. Active in Florence, Italy, at the close of the fifteenth century, he is often called the Renaissance Man, and the Universal Genius. Five hundred years later it is very difficult to fathom his logical manner of reasoning and acting well beyond what was the most advanced science and art in the world at that time. It would be most useful to somehow find ways to educate aspiring youth to emulate the most unusual mindset that he possessed.

These interesting examples being envisaged as mergers of art and science can aid progress in several ways. One is to motivate pride of accomplishment within someone who produces a new item that not only functions well but also appears to the public as pleasing to the eye. Another is to substantially improve marketability when the public is bedazzled by its physical attractiveness. Portraying opportunities to design composite art/technical projects can boost an existing staff to approach challenges in a refreshing manner and greatly aid recruiting new talent. Perhaps most importantly, professional educators could stimulate new students to become interested in joining the scientific and technical community: if they are already attracted toward art, here is an opportunity to pursue that interest while also developing a miraculous piece of machinery to improve people's lives. The national need for new participants in all technical fields is growing and never ending.

One attempt to fathom the way inventors think and react, as well as to intrigue and attract new participants in science, is expressed in the book *Lab Girl* by Hope Jahren. It calls attention to one's meaningful internal motivator to learn and to attempt new adventures: curiosity.

> Maybe I can convince you. I look at an awful lot of leaves. I look at them and I ask questions. I start by looking at the color: Exactly what shade of green? Top different from the bottom? Center different from the edges? And what about the edges? Smooth? Toothed? How hydrated is the leaf? Limp? Wrinkled? Flush? What is the angle between the leaf and the stem? How big is the leaf? Bigger than my hand? Smaller than my fingernail? Edible? Toxic? How much sun does it get? How often does the rain hit it? Sick? Healthy? Important? Irrelevant? Alive? Why?
>
> Now *you* ask a question about *your* leaf.
>
> Guess what? You are now a scientist. People will tell you that

you have to know math to be a scientist, or physics or chemistry. They're wrong. That's like saying you have to know how to knit to be a housewife, or that you have to know Latin to study the Bible. Sure, it helps, there will be time for that. What comes first is a question, and you're already there. It's not really as involved as people make it out to be.

2

ELECTRIFYING DISCOVERIES

THE PREVIOUS ILLUSTRATIONS occurred in very different working environments: some were only one-time breakthroughs, some an evolving set of improvements, and some as ways to solve knotty problems. There are numerous other marvels that emerged after many years of sustained effort by individuals or teams of inspired technical thinkers. In this chapter are some extremely useful devices that employ the wonders of electricity. Each description will begin with the likely motivating rationale that inspired so much dedicated effort to arrive at success. Herein we focus our search for inventiveness on the mysteries of electromagnetism. This physical phenomenon could be considered the mainspring basis for the best technical advancements that benefit human lifestyle. It can be useful to speculate on the thought processes used in creating these devices so we may emulate those as we attempt to inspire others.

The public usually credits three of these to have been conceived by singular inventors: the light bulb (Edison), the telephone (Bell), and the radio (Marconi). That is because each of these inventors was the first to obtain patent rights to reproduce and market their sought-after products. But it seems more appropriate to credit the entire array of goal-driven people whose accumulated successes were the real foundation for the final optimal implementation. It is almost impossible to write a fully comprehensive credit list, but the foremost contributors are historically recorded. This chapter's focus is on the history of these notable inventions, showing how each involved considerable evolution by many talented inventors and engineers.

A fine example of composite skills combining to formulate and clearly understand and define the magic wonders of electromagnetics occurred in the nineteenth century. Englishman Michael Faraday discovered through diverse experimentation that electrical waves always have inter-coupled magnetic waves. Visualize an electrical wave as a ribbon streaming by with its wider part vertical to your view. Linked with it is a ribbon with its waveforms horizontal; this is the magnetic wave, forever proceeding as an energy mate. These linked ribbons zip through space at the speed of light, 186,000 miles per second. Faraday exploited this phenomenon by finding a way to employ electricity to drive mechanical power devices. Today's electric motors spin because electrical power excitation is rapidly switched between succeeding magnetic segments placed in encircling rings. With thorough understanding of Faraday's experimental confirmations, Scotland's James Clark Maxwell evolved the mathematical means of defining and predicting the behavior of these energy waves throughout the entire frequency spectrum. Those calculating tools are still used today and have paved the way for the myriads of applications extant. Other than worldwide distribution of public electrical power, the greatest electromagnetic gifts to improve human lifestyle, in the author's opinion, are wireless communication and computers.

Let There Be Light

It is easy to understand the urge to perfect a continuing light source. We all would like to be able to function in darkness and inclement weather, but only crude means had been found over centuries of attempts. Great commercial benefits could accrue to the inventors. Electrically powered **light bulbs** are usually credited to Thomas Edison, who in 1879 was granted the first patent protection for an incandescent lighting device. Essential components were a glass vacuum enclosure, a pair of electrical conductors, and a filament stretched between them.

The first known attempt to create such a device was made in 1802 by England's Humphrey Davy. During the next seventy-seven years, before Edison's patent, nineteen other scientists made meaningful additions to the art and science of light bulbs. The principal difficulty was finding an economical material for the filament that would sustain the heat during incandescence, tolerate switching surges, and last for a long time. Experimenters used fiber threads of carbon, platinum, carbonized paper, cotton, linen, and strips of various woods. Edison's success resulted from three significant improvements: a deeper and sustainable vacuum, a dependable electrical power source, and a long-life filament material. The latter was found after numerous searches: a carbonized thread of Japanese bamboo.

Edison's very notable advisory phrase was, "I have not failed. I've just found 10,000 ways that won't work."

This is certainly a meaningful guideline for others trying to do something new: expect and accept unsuccessful trials, and persist until desired goals have been met. It is astonishing today that in Livermore, California, a carbon filament light bulb has been electrically powered and giving light since 1909.

Edison Light Bulb

Neighborly Conversation

Similarly to the drive for a light-bulb market, there was great potential for selling a device to permit personal communication between remote sites. Development of the **telephone** was an even more complex amalgam of discoveries and might also be considered as a logical progression of the landline telegraph system. The first commercial telegraph network, assembled by England's William Cooke and Charles Wheatstone, began operating in 1837 in the five-mile span between two villages. In a few decades, with many further improvements, a plethora of land and undersea wires could be found throughout the world. It was the most significant rapid long-distance communication method used by both sides in our nation's Civil War. By that era, information in words was transferred into Morse code dots and dashes keyed with a switch by a sending operator. (This code was patented by America's Samuel Morse and Alfred Vail, becoming the world standard in 1844.). Then the resulting spaced on-off electrical pulses passed along the wires strung on poles, activated a sound-clicker at the receiving site, and were deciphered back into words by a receiving operator. As many as 90 words per minute could be sent in this manner. By 1874, Denmark's Poul la Cour—and separately in America, Elisha Gray—found that placing ringing tuning forks or steel reeds within a magnetic field generated matching electrical current vibrations that could be sent along a wire to a receiving site. This allowed "multiplexing," wherein several different messages could be sent at different tone levels at the same time. Commercial traffic could then use a single telegraph wire line to send seventy-two independent message streams. The receiving station was equipped to separate messages by adjusting to the desired tone. Tuning fork experiments also represented the first electrical transmission of audio tones. What was now needed was a reliable method to accurately and reliably convert voice or complex musical sound waves into electrical vibrating streams that could be transferred long distances by wire.

For centuries, children used a simple toy to speak in quiet tones across a room. Paper or metal cups were connected at each end by a taut string. Spoken words at one end generated vibrating air waves that pushed on the diaphragm spanning the cup's bottom, making it synchronously vibrate. These motions vibrated the tight cord, in turn causing the receiving diaphragm to wiggle. That motion shook the air within that cup to form sound waves; the recipient could then hear those same words. How to convert those diaphragm motions into electrical signals and assure that they were of sufficient power to carry long distances?

Tin Can Telephone

Attempts to convert these mechanical motions to electromagnetic waves were begun in 1833 by experimenters in Germany, France, England, Italy, and the United States. Most significant was the conceptual and experimental work done by Italian Antonio Meucci as early as 1834. He filed a patent claim for the electromagnetic telephone in 1871 and became the chief rival to the eventual 1876 patent awardee, Alexander Graham Bell, a Scotsman who had migrated to Canada, and then to the United States. This long-lasting argument about "who was first" became a political dispute between Canada and the United States. Bell's big breakthrough was evolving a smooth conversion of the air sound waves into a continuous electromagnetic stream. Although Bell's first patent used a derivative of American Elisha Gray's liquid-form converter, Bell's 1877 patent described one that was far more producible in practical quantities. The transmitting site used a movable iron rod pinned to the center of a circular skin diaphragm. Vibrating air pressure from spoken words funneled to that membrane by a conical mouthpiece moved the iron rod, inducing electrical waves. These passed along the wire pair to a receiver. There, a soft iron membrane at one end of a metal cylinder vibrated when excited by that current passing a surrounding coil, producing sound waves in nearby air.

Freedom from Wires

It took decades and the efforts of many teams to reach where we stand today in **radio** communication. Enormous market potential beckoned those who could produce products that achieved data transfer over long distances through the air. Funding support was from scientific supporters, private investors, and governments. The goal and powerful drive of many scientists was to find a way to rapidly communicate without the burden or limitations of stringing wires between sender and receiving sites. There were five significant technology thrusts that made over-the-horizon linkage truly effective: (1) spark gap transmission; (2) vacuum tube devices; (3) carrier wave modulation for information clarity and sensitive microphones; (4) transistor and solid-state components; and (5) digital conversion along with micro-miniaturization.

Based on Maxwell's electromagnetic theory; experiments of Britain's David Hughes in 1878; and efforts in the 1890s by German Heinrich Hertz, Serbian-American Nikola Tesla, Spaniard Julio Cervera Baviera, German Karl Ferdinand Braun, and Italian Guglielmo Marconi, radio communication became possible using **spark-gap transmitters**. This signal source method became the principal basis for wireless telegraphy until the 1920s.

When electrical high voltage is applied across a small gap between two electrodes, a spark results, creating radio-frequency electromagnetic energy. This is captured and passed to an antenna that radiates the signal skyward. At some distance away, an antenna with the right shape will pick up these waves and convert them to electrical currents. Such transmitters produced radiation over a wide band of frequencies, requiring special placement and timing to avoid interference from unwanted sources. Information is inserted into the transmitter by switching the power on and off, forming Morse code dashes and dots, just as was done in landline telegraph networks. Messages were activated by a radio operator, commonly known as Sparks.

Spark-gap Transmitter
(photograph: Electrical Simplified)

Vacuum tubes appeared in the 1910s. These most capable devices stemmed from experiments in 1873 by England's Frederick Guthrie, followed by many other scientists, and were finally patented in 1884 by Thomas Edison. By 1907, Britain's John Fleming, Austria's Robert von Lieben, and America's Lee de Forest perfected this component that led the way to enabling almost unlimited functions to be performed by electronics. Substantial radio benefits included: transmitting in continuous waves instead of pulses; operating many stations at the same time by controlled frequency separation; the ability to perform throughout the full radio carrier frequency band (148 KHz to 3 GHz); very crisp on-off actions when desired; power-usage efficiency by employing only single-side-band (SSB: only the top half or positive portion of the radio wave is used); and excellent ways to inject information in a message stream. Additionally, vacuum tubes solved the inability to communicate over the horizon. In many parts of the radio frequency band, adequate reception may require line-of-sight between the sending and receiving antennas. However, 1.6 to 3 MHz is termed **short wave.**

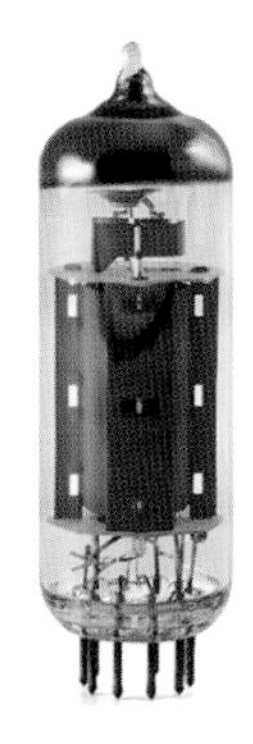

Electronic Radio Tube

Signals sent in this band have the magic property of bouncing or "skipping" off the Earth's surrounding ionosphere. Reflected energy waves allow reception well beyond the visual horizon limits. Shortwave was very effective on December 7, 1941, when the author's father, working for the Mackay Radio & Telegraph Company, sent the public news stories of the Japanese attack by Morse Code from Honolulu to San Francisco at a rate of 90 words per minute. This lasted only twenty minutes; the Navy Shore Patrol arrived and shut the station down to avoid inadvertently sending classified military secrets.

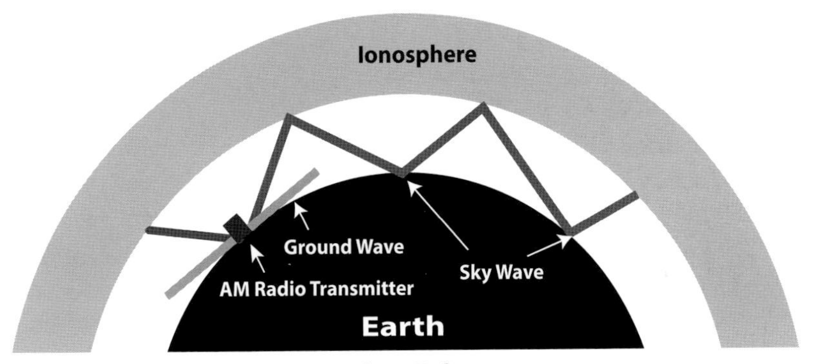

Ionospheric Refraction

Early designs of **microphones** used diaphragms impinging on carbon granules to convert air pressure of sound waves into electrical signals. Demonstrated in the early 1870s by England's David Hughes, a U.S. patent was awarded in 1877 to America's Thomas Edison (contested for years by German-born American inventor Emile Berliner). Better clarity designs of several types emerged from 1916 to 1930, the most notable being done by E.C. Wernte, H.J. Round, and Alan Blumlein in England.

Vintage Microphone

By 1930, modern "mikes" were able to transform high quality music into precise electrical waveforms.

Public radio broadcasting began in the 1920s. The wonders of clearly sending voice and music across great distances was done by continually altering the transmitted carrier waves to implant intelligible information. This is called **signal modulation** and is performed in several ways: on/off for Morse and digital codes, amplitude modulation (AM) for voice and music, and frequency modulation (FM) for finer grade music. In AM, incoming information alters the power level, which you can visualize as wavy lines atop the carrier radio waves. The receiver can extract this information and a speaker instrument converts it back to sound waves identical to the input.

Amplitude modulation was first tried in 1906 by Canadian Reginald Fessenden, using a spinning disc that varied the signal amplitude 10,000 samples per second based on voice inputs from a microphone. The output was garbled; later use of vacuum tubes with continuous carrier waves made AM quite clear. Greater fidelity of the sound track can be achieved by using FM. In this case, the carrier frequency itself is altered by inputting the microphone's output. Since music sound waves are usually less than 20 KHz, the variation of FM's operating 88 to 108 MHz carrier represents only a small change. Precision is greatly enhanced, but line-of-sight between sending and receiving antennas is required for this high frequency band, causing short-range limits, a bit inhibiting for the commercial radio market.

Vacuum tubes did their job quite well but suffered from being fragile, generating much heat, being heavy and costly, and often unreliable or short-lived. Patents for a **solid-state triode** were filed in 1926 by Austrian-American Julius Lilienfeld and in 1934 by German Oskar Heil. It took until the 1960s to obtain pure semiconductor materials to mature such components. Initially, silicon and germanium were used; their electrical conductivity can be varied by imposing an electromagnetic field, performing functions similar to those of the gridded vacuum tube. In 1947, America's John Bardeen, Walter Brattain, and William Shockley finally achieved full success with the transistor. This began the dramatic expansion of electronics into every possible usage.

Since the 1970s, great leaps in performance stemmed from **digitizing data**, in which information of frequency, phase, and amplitude are converted to binary numbers. These could be manipulated mathematically for time compression or for combination with other data to yield a more comprehensive conclusion. Also, devices could now be easily managed by programmed software commands.

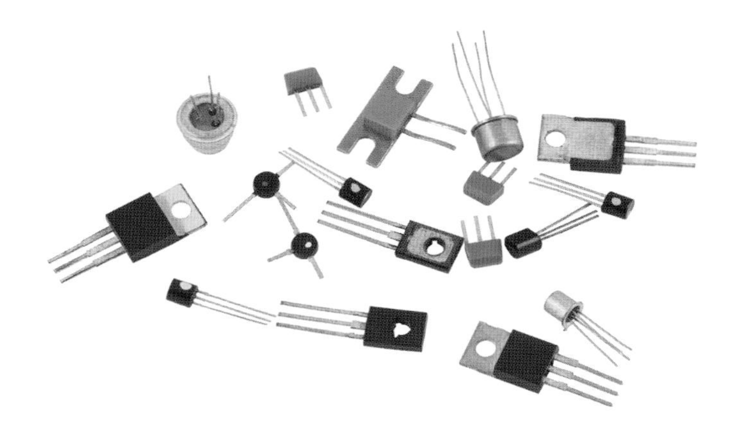

Transistors

Microminiaturization (placing thousands of electronic elements on wafers bonded into a tiny chip) yielded functional complexity at low cost, high reliability with long life, and incredible operating speeds. Benefits to human communication from these two dramatic changes included better clarity and fidelity, the ability to transfer many signal streams at the same time using only a small frequency band, and the ability to transfer masses of complex data in a short time.

Moving Sights from Different Sites

There was an enormous business-growth lure for consumer products that now added to that magic of radio with real-time pictorial motion. That was a big incentive for huge corporate and government investment. Most inhabitants in the world these days have access to **television** (TV), a system for broadcasting dynamic images to remote sites. These operate in the electromagnetic bands called VHF (very high frequency) and UHF (ultra high frequency), ranging from 30 MHz to 3 GHz. The signals are imposed on the broadcast carrier frequency by AM, using single-side-band (see above discussion of radio). The audio signals are done using FM. Reception units may obtain broadcast data streams directly with line-of-sight antennas, via transmission lines from central distributors, or through satellite relay linkage.

Television Screens

Major elements of color television are a video camera to render electronic data streams of images being observed, an amplifier of the resulting electrical signals, processing to set format, linkage to a distribution system, and a receiver feeding a display tube. Until the digital era, images were formed by narrowly spaced horizontal lines across the picture width. Completed scans of each "raster" are made 30 times per second, a rate fast enough that the human eye cannot perceive the scan's motion, and the lines are close enough so as to blend as a solid picture. Picture density is set by the number of lines in each image. The first video raster was demonstrated by Japan's Kenjiro Takayanagi in 1926; it used 40 lines. In 1941, the United States set 525 as its imaging standard (for processing efficiency, only 480 were displayed). The Soviet Union selected 625 in 1948, which was soon adopted as the European standard raster. As the beam scans each line, it detects levels of light-intensity. To display what the video camera has seen, a signal stream of these samples is sent to the display cathode-ray tube (CRT), described below.

There were many people throughout the world endeavoring to piece together the numerous complexities of widespread television broadcasting. The first device to scan images was the facsimile machine made by Scotsman Alexander Bain in the mid 1840s. Many other image devices emerged from Britain, Japan, Germany, the United States, and Russia. These all used mechanical scan methods, and all were limited in amplitude for transmission and poor in showing moving subjects. Finally, with the appearance of useful iconoscopes and CRTs, in 1934 Philo Farnsworth of the United States achieved the first all-electronic live-scan.

Sir Isaac Shoenberg, Russian-born British electronic engineer, of EMI (Electrical and Musical Industries) in 1936 showed the first "high definition" video with 405 scan lines (this term today has far more meaning with resolution using digital processing and plasma or LED displays). In 1937, the BBC showed the first outside live filming; and the Philo Farnsworth system was successfully used for public showings during the 1939 New York World's Fair. The first color broadcasting demonstration, achieved in 1951, was seen by the public in viewing auditoriums in only five U.S. cities. For many reasons, widespread networking and household viewing took another fourteen years.

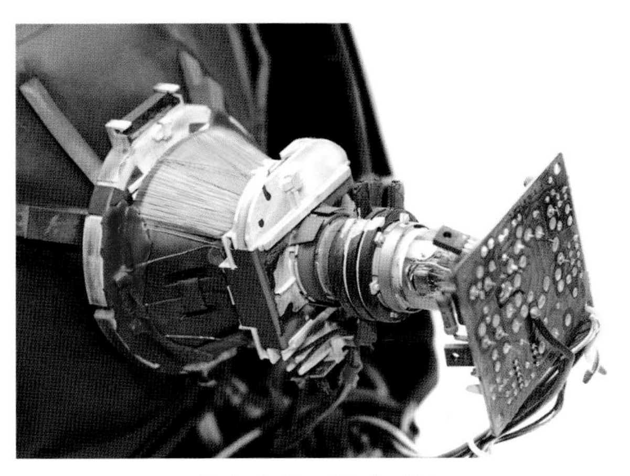

Cathode Ray Display Tube

Of the two major devices needed for television, the display tube was first to emerge. The **cathode-ray tube** has a conical-shaped glass vacuum tube with a large faceplate or **viewing screen** at one end and a small cylinder at the other. A uniform coating of phosphorescent material covers the inside of the faceplate. The cylinder houses a gun that will create a beam of electrons to be steered at the phosphor coat. The beam is formed by electrons escaping from a tiny post in the cylinder (the cathode) when differing voltages are applied. Deflection of the beam for pre-programmed scanning is controlled by electromagnetic coils surrounding the gun; circuitry guides the beam in successive sweeps forming the raster. As the electrons hit the phosphor (the anode), each spot is stimulated to emit light that can be viewed on the CRT's outer faceplate. Brightness of each spot has been set by the energy level for that specific location as it had been sampled by the iconoscope's scanner. Black-and-white displays have one gun, and phosphor coatings are usually copper or silver activated zinc-sulfide. Color displays use three guns; the faceplate coating is a mosaic array of phosphor particle materials that emit either red, blue, or yellow.

Those differing materials are interspersed in the faceplate coating so that every color dot is directly next to the other two. Each gun beam is matched to one color and is moved so as to stimulate only that particular color dot. The mosaic spots are so closely spaced that the human eye will blend together emissions from adjacent particles. Differing stimulation levels of each will result in hues from any part of the rainbow.

The first CRT was demonstrated by German Ferdinand Brain in 1897, using the 1869 findings of German Johann Hittorf, as well as Britain's Arthur Somerset and William Crookes. The first commercial product emerged in the United States, designed by John Johnson and Harry Weinhart, and then produced and marketed in 1922 by Western Electric Company.

It took many years to mature the other key element needed for television: an iconoscope or **image-scanning device** that would convert a visual image into a stream of electrical signals that could be sent by radio waves to a television receiver. Efforts were actively underway in the United States, Great Britain, Germany, Hungary, Russia, and Japan. Paul Gottlieb Nipkow of Prussia patented in 1883 the Nipkow Disk.

Electronic Image Scanner

It had many equally spaced tiny holes pierced in a spiral pattern from its circumference to its center. A lens placed an image upon the disk face. As the disk was rapidly spun, light passing through each hole traced a line across that image. Each hole in that spiral traced another raster scan line, spaced by their differing distances from the spin center. Beneath the disk were photo-sensors that formed electrical signals, to be passed to a receiving device. An identical disk there, with spin exactly synchronized to the transmitting disk, would then project matching light lines onto a display. When perfected, this became the first mechanical scanner and was used by most TV experimenters until 1920. Especially successful was the invention by Hungarian Kálmán Tihanyi in 1926. This new creation enabled successive sets of photons to be accumulated by charge-storage.

This concept was further improved by Vladimir Zworykin in the United States. By 1933, the device was called by RCA the "iconoscope," and became the principal video camera in our country until 1946, then replaced with a far-superior *orthicon*.

Another excellent camera was perfected by Germany's Telefunken; it was used to demonstrate TV during the 1936 Berlin Olympic Games and soon became the standard for European TV. Their moniker of the device can be relished by those who are a bit ignorant in the German language: *Lichtelektrische Bildzelegerrohre für Fernseher* (photoelectric image dissector tube for television). The British design, made by EMI, was dubbed "the super-Emitron." Later cameras used for many years were called *vidicams*.

The conceptual configuration of these cameras were cylinders with an optical lens at one end to peer at a scene, and a three-layer thin sheet closing the opposite end. That diaphragm's lower layer was a metal mesh; aluminum oxide insulation formed the center layer; atop that and facing the lens was a large array of carefully spaced photoelectric granules of potassium hydride. Just inside the lens but out of view was an electron gun. It cast a beam in a scan pattern, exactly timed to match the CRT in the receiving TV. The raster was traced once, which, when striking each granule, created an electrical charge between it and the lower mesh screen. (This acted as a storage capacitor for that spot.) If there is no incident light, all these charges will be equal. When struck by an incoming image, the photoconductive granule will have additional electrons stored; granules hit with a brighter bit of the image will accrue more than those hit by a darker part. For readout, the beam scans each granule again: those with more electrons will not be recharged by the beam. Any recharge resistance is reflected to a circular collecting ring. Those sequential granule sample responses form an electrical signal for that spot. That signal stream is then amplified and used for broadcasting.

Fabulous leaps in image quality became possible in the twenty-first century. Transmissions used **digital data streams** instead of analog; displays became flat-screens using plasma cells, liquid crystals, or light-emitting diodes (LEDs). These also have full-frame memory storage. Rather than using raster scans to form images, picture elements on these displays are placed with a two-dimensional position grid locator. These spots, or picture elements, were renamed pixels after everything went digital. The word was coined in 1965 to describe the image elements being received from a camera recording space observations. Television equipment is being employed currently for some spectacular uses. Some of these are: education of students nationwide from a single teacher; space probes showing the surfaces of Mars and Saturn (the first distant images of Earth were viewed by the Hughes Surveyor after its soft landing on the Moon and sent to home screens in 1963); examination of the human heart using a tiny electronic viewer at the end of a catheter inserted by way of a leg artery; views of internal structures using x-ray images; and giant stadium screens for attendees to see full-speed or slow-motion replays of sports events. Home and hotel entertainment now can be watched as high-definition images on 72-inch or larger big-screens. Hundreds of channels can now be received at home by way of satellite relay networks. The initial demonstration of TV linkage between continents was a live-broadcast across the Pacific Ocean of events at the 1964 Tokyo Olympic Games. It was effected by a relay network using the Hughes *Syncon* geosynchronous satellite, a dramatic sign of things to come from operating from space. Additionally, perfection of lasers and fiber optics enabled high-capacity networks using undersea cables.

Overcoming Foul Weather and Darkness

Systems using the next higher frequency bands are **radars**, an acronym for "radio detection and ranging." Perfecting this type of system was stimulated by the urgencies of national survival in two world wars, so copious budget allocations provided by many nations assured success. Operating frequencies are categorized into eight bands between 3 and 300 GHz. Bands are identified by code letters such as L, C, and X. Each of these can yield differing performance variables, including power level, resolution,

Air Traffic Control Radar

and physical size. Radars are now an essential part of our high-technology lifestyle. Numerous applications include air traffic control, satellite communication, precision global mapping, navigation at sea, collision warning for automobiles, and even a variant in microwave ovens.

Visibility is often limited by the earth's horizon, darkness, adverse weather, and the object's size. These constraints can be conquered by this electronic wonder when properly located. After war broke out in 1939, large and very heavy rudimentary radars were installed on ships and at ground bases. Active development in Germany and Britain greatly improved performance. Size and weight reductions plus a change to higher operating frequencies enabled installation in aircraft for navigation, ground observation, and target detection. British projects carried colorful names: Ash, Boozer, Chain Home, H2S, Monica, Perfectos, and Village Inn. The Germans used Berlin, Bremenanlafe, Hohentwiel, Lichtenstein, and Neptun. The United States also had rudimentary ground and shipborne radars for detecting incoming aircraft raids. One set was operating in 1941 on the north coast of Oahu; its two operators spotted the hundreds-strong Japanese raid heading to Pearl Harbor. Unfortunately, their superiors did not believe their telephone report, either thinking their subordinates too inexperienced or presuming that the radar blips were a group of friendly B-17 bombers known to be en route from California.

Today, radars consist of many hardware segments that have different functions. The segments are electrically linked together to operate as an integrated system: a frequency source governing the radar's operating signals; a transmitter that amplifies the output; an antenna to project a shaped beam into the area being examined as well as capturing echoes reflected from objects ahead; a receiver; and a signal processor to interpret the echoes. The data is then fed to a computer that manages the system as well as the controls and displays for the operator. During the 1940s, airborne radar transmitters used a cavity-magnetron tube to provide carrier frequency and as the power source. Those radars projected a continual series of short power bursts at low-PRF (low pulse repetition frequency) between 250 and 2,000 times per second. This allowed echoes from targets up to fifty miles distant to be seen between outgoing pulses. Pointing the main beam at the ground caused background returns one thousand times the echo strength from small objects. This huge ground clutter totally obscured many desirable airborne objects that reflected only low-level echoes.

Continual drive for increased performance and immunity from operating limits spurred many new design concepts. Advances in the 1960s overcame the clutter obscuration problem by exploiting the

Doppler effect, first physically measured in 1842 by Christian Doppler, an Austrian doing research in Prague, Czechoslovakia. Everyone who has heard a moving train's whistle has experienced this: as a train approaches, we hear a higher pitch sound, then we hear a lower pitch as it speeds away. When hearing an incoming train from a fixed position, the sound waves in the air are compressed, making the vibrations closer together, therefore, at a higher frequency (higher pitch). The reverse happens as the train departs: sound waves are stretched to a lower frequency. A radar having a precise output frequency can determine shifts in return echo frequencies. These are exactly proportional to the comparative speeds of its mother aircraft and the flying object being observed. This measurable shift sharply separates the desired echoes from that dreaded heavy clutter echoed from the unmoving ground. Essential to making a radar perform in Doppler is to assure a very stable transmitted carrier frequency. This is not possible using the unstable magnetron of the old days.

At Hughes Aircraft Company, Drs. John Mendel, Bill Christoffers, and Mal Currie perfected a lightweight, highly stable power amplifier called a **traveling wave tube** (TWT). This was a derivation of a device first constructed at a British Admiralty radar laboratory and later improved at Bell Laboratories. It establishes an electron beam in a long cylindrical vacuum tube to amplify a microwave signal emanating from a frequency-precise oscillator. The beam is guided through a series of lightweight doughnut-holed magnet wafers to a waveguide leading to the antenna. The amplification proceeds without altering the oscillator carrier frequency. This finally enabled the precision needed to perform Doppler shift detection and measurement.

These radars used high pulse repetition frequencies up to 25,000 per second rather than the low-PRFs of 2,000 described previously. This enabled use of very high average power to greatly extend detection

Traveling Wave Tube (TWT)
(courtesy of Raytheon)

and tracking ranges. However, special processing was implemented to overcome data confusion: many outgoing pulses were sent before a long-range target reflection had bounced back. This component breakthrough fostered patent protection that set up a business bonanza for Hughes; consequently, Hughes enjoyed many years as the world's largest producer of many different types of TWTs.

There are many varieties of radar antennas, depending on the mission to be fulfilled. Hughes was at the forefront of their technology evolution: conical scan reflectors; space-borne pattern-shaping horns; gimbal-mounted planar arrays; beam scanning in all dimensions without physical motion; and arrays of separate elements, each actively transmitting and receiving signals. Very significant was our creation of **electronic scan** replacing that slewing by hardware motion. Mechanical scanners require heavy mounting support, usually are placed on complex gimbals for motion stability, can require large driving power, and may be slow in slewing the beam. In contrast, shaping and pointing the beam electronically can be done instantly, several beams can be formed and used simultaneously, and the support structure can be quite simple.

During the 1960s, Drs. Nick Begovich and Nick Yaru led an effort that perfected the world's initial electronic scan antenna at Hughes. The first method was to feed the signal continuously to lines of antenna outlet elements while slightly altering the carrier frequency. This caused the whole antenna to function like a visual prism: white light entering one facet of that glass wedge exits in separate colors. The frequency of each color differs, which alters the travel distance needed to go through the glass, thus changing the emergence angle from the slanted exiting edge.

Phase scan soon became a better way to steer the beam. The phenomenon was described to me by Dr. Art Chester. Visualize dropping a pebble into a quiet pond, starting a circular ripple that spreads equally in all directions. If several friends also do the same, each pebble makes its own circular wave, creating a jumble of confusing wavelets. If everyone makes sure to drop the pebbles in a straight line at the same moment, instead of a confusing pattern of ripples, you will see a smooth blended wave moving across the pond, with the front of the wave parallel to the line of splashes. All the circular ripples have apparently joined in one direction to make a wave-front line. Dropping the pebbles one at a time in a timed sequence from left to right will point the wave front in a different direction. Instead of being parallel to the pebble drop line, the combined wavefront will move to the right: the ripple from the first pebble arrives at the second ripple just as that pebble hits the water. This dynamic event continues at each following pebble splash. By adjusting the timing, we can produce a plane wave traveling in whichever direction we choose.

When using any electronic signal, it is most helpful to know not only its frequency and its strength but also the arrival time of each part of the wave. This latter is called "phase." If you watch a series of ocean waves approaching the coast, you will see that the crests and troughs are at approximately equally spaced distances. The crest's peak represents the wave's strength (amplitude). If the sand below is flat, halfway between the top and bottom of the wave the water is at the same level as a placid sea. Specific points (phases) of the wave can be numbered in degrees just like positions on a circle. As the wave approaches, the first point at sea level is defined as 0 degrees; the crest is 90 degrees. As the wave surface moves down toward the dip, it passes sea level again (180 degrees), and the trough is 270 degrees. Then the next wave set repeats this sequence. So, every degree of phase defines an exact position in time and amplitude in the wave's surface.

Hughes Air Defense Radar
(courtesy of UNLV)

For our new radar let's compose the antenna as a number of small controllable apertures and send out simultaneous signals so that all the waves add up in the desired pointing direction. Similar to the pebble game, a group of antenna apertures can do the same kind of "beam steering" by adjusting the time delay from one small antenna outlet to the next one. Since a time-delay moves to a later point of the signal wave's up and down cycle, we call it a "phase delay," and the whole beam control process is called a "phase scan." A longer time interval means a larger tilt angle. This phenomenon can be done whether transmitting or receiving.

And There Is More

In chapter 11, many more electromagnetic marvels will be discussed. All those complex devices have greatly improved human lifestyle for the last fifty years. Included are exploitations of the next higher frequency portions of the spectrum: infrared and visible light. Beyond that are ultraviolet, x-rays, and gamma rays, which reach beyond the scope of this book.

3

PROVIDING PARADISE FOR CREATIVITY

WHEN AN INDIVIDUAL or a group of people are motivated to establish a large corporation to achieve complex goals, there are many strategic and tactical decisions to be made. It will not be easy to "do it right" since so many uncertainties lie ahead. The planners and implementers must deal with interactive factors and be ready and willing to alter course when roadblocks arise. Basic essentials are selecting a location, securing appropriate facilities, forecasting staffing needs, planning recruiting methods, and assuring adequate funding sources. Then the tricky things begin: how best to organize and manage the workforce, to keep them motivated and productive, to assure the market opportunities can be met, and that competition threats can be handled. This chapter's content and that of the next one are intended to meld bits of history with experiences in the successful aerospace industry. Such an amalgam should be effective in meeting today's dynamic technology advances. The godparents of the new enterprise will be aided in making well-founded decisions on these factors by absorbing the tenets and wisdom described in the following five noteworthy publications about management style (see appendix for details).

The Essential Drucker by Peter Drucker
How to Win Friends and Influence People by Dale Carnegie
My Years at General Motors by Alfred P. Sloan Jr.
The One-Minute Manager by Kenneth Blanchard and Spencer Johnson
The Six Sigma Way: How GE, Motorola, and Other Top Companies are Honing Their Performance by Peter Pande, Robert Neuman, and Roland Cavanagh

One real challenge in today's organizations is to provide a stimulating facility and motivational structure that will assure a continuing supply of useful ideas. Most working professionals are strongly influenced by the working environment surrounding them. If an employer can provide a comfortable and pleasant ambience, positive motivational results will be realized. Inspiring performance excellence may also be spurred by opportunities for attaining better office quarters as well as by the usual reward of upward growth in responsibilities. Persistent drive to accomplish new goals also can be influenced by the geographical placement of the workplace. History seems to indicate that tropical climates may foster a casual attitude, since life can be easily sustained.

World Climate Zones

When colder climes are the norm, people feel a more urgent need for productive efforts to assure a comfortable existence. Right in between these extremes, outfits in the temperate and semi-tropical zones can enjoy the benefits of environmental bliss, together with the drive to gain rich compensation and prestige rewards for continual creative accomplishments. Dr. Art Chester, former President of HRL Laboratories (the central research lab for Hughes Electronics), holds strong views regarding attracting employees with technical creativity:

> Talented engineers are dominantly motivated to solve problems. Management must present them with a legitimate need or technical difficulty to be overcome, and be prepared to negotiate its details. A daunting problem without a current solution is a tremendous driver to the kind of person who studies

engineering or science. We still need a positive working environment, and that helps to promote creativity. However, first must come presentation of worthwhile things to remedy, otherwise the creativity that emerges will be useless, harebrained ideas. This element is as important as amenity factors, and logically comes first in management priority.

Climatic Influence

Numerous factors play a role in selecting the optimum location for a new facility intended to culture innovation. These include cost of living, tax burdens, educational opportunities, transportation, safe living conditions, and helpful government surroundings. However, a huge factor to attract a happy staff is regional climate.

Being centered in the Southern California culture and climate certainly helped the aerospace industry to dominate that field in the United States. The area first expanded economically by phases: cattle ranching, orchard farms, exploiting rich oil fields, then becoming the world center of the movie business, and then leading in aircraft development. Famous firms emerged: Douglas, Convair, Lockheed, North American, Northrop, Rockwell, and Ryan. All benefited in recruiting employees and keeping their supporting families happy by locating in this comfortable climate. Remarkable were the World War II high-performance fighters, the Lockheed P-38 *Lightning* and the North America P-51 *Mustang*. Also, the Douglas C-47 *Skytrain* became the Allies' transport workhorse. These firms and the suppliers employed about a million workers during the very active years around 1980. And all played a significant contribution to the United States "winning" the forty-five-year Cold War against the Soviet Union. That was done without any combat, since the Soviets experienced virtual economic collapse trying to match the superlative performance of our weapon systems. Well known are many fine commercial airliners, including the Lockheed *Constellation*, Douglas *DC-10*, and Convair *660*.

Another stalwart was Hughes Aircraft Company, established in Burbank in 1932. After moving to Culver City in 1941, it started its push in 1946 to eventually become the world leader in defense electronics. The headquarters near Los Angeles was ideal: one mile from its facilities were the Pacific beaches, a mile the other way was an RKO filming studio lot, and two-hours' distant were several winter skiing resorts. The magnificent Sierra Nevada range, with peaks soaring to over 14,000 feet, was a magnet for camping or backpacking; recreational boating marinas were nearby; magic Santa Catalina Island beckoned; three major universities were only a half-hour's drive; and the weather was close to perfect. Staff members could plan to watch the annual Rose Parade in Pasadena and afterwards cheer or boo the opponents at the celebrated Rose Bowl football game. They might even run into a famous movie star at a restaurant on the fabled Sunset Strip. Many apartment rentals and affordable homes were within a short commute. Who could ask for anything more? Recruitment from the eastern and central states was usually a piece of cake.

In addition to this attractive ambience, the frenetic expanse of electronics evolutionary challenges, attractive compensation levels, and the aura of mystery that surrounded Howard Hughes were powerful lures for potential staff members to Hughes Aircraft Company. Recruits came from all parts of the country and included many professors leaving their university posts for a more fruitful career. These factors, and a growing reputation for professional excellence, helped attract the best talent, eventually leading to 22,000 engineers, including 4,000 PhDs. With that array of brainpower, anything became possible, and continuous imaginative creations sprang forth.

Setting the Stage

Once the central geography has been set, the corporate leaders must plan and provide the proper **vibrant workplace**. For technical creativity, it has been found that the facilities should resemble and operate like a laboratory. The Apple computer dictionary simply defines that as a "room or building equipped for scientific experiments, research, or teaching, or for the manufacture of drugs or chemicals." What seems most important when considering innovation in electronics is the emotional aura embodied in the words "research" and "experiments." These imply exploring the unknown, forming physical and mathematical theories to reach in new directions, and accepting trial-and-error learning in order to garner progress. The imbedded philosophy of staff and management is vital and can be enhanced by the appearance of the working environment. This usually means workspaces for quiet thought and nearby spaces and equipment for hands-on experimenting with prototype hardware. Everywhere the spirit of free-thinking abounds.

Over the last five decades, there have been periodic changes in attitude about what constitutes an ideal office layout. To accommodate significant clusters of employees doing repetitive tasks—such as in mass-production manufacturing—large open spaces are ideal. Facility investment and upkeep costs are low, processing steps can be laid in geographic sequence to limit subassembly transportation, and supervisorial visibility is comprehensive. For clerical and similar tasks, accommodations can focus on employee comfort and access to storage space. For financial and legal professionals, attention to information security, computer access, adequate filing space, and relative quiet are appropriate.

Workplace Nirvana

The perceived best office layout style for engineers and scientists seems to oscillate each decade. Aerospace firms during the mid–twentieth century usually placed large numbers of design engineers in "bull pens," large open spaces filled with activity and noise. This configuration not only was frugal in building construction and upkeep but also was believed to promote teamwork and conviviality. This open-space arrangement is favored today by many creative organizations in order to ease cross-fertilization of ideas and expertise, particularly in the entertainment professions. Hughes Aircraft very much favored the privacy style, with individuals sharing workspace rooms with only one to three office teammates. This arrangement assured quiet for intensive thought and also was perceived by employees as a form of recognition for their importance in the company. This marked a sharp contrast to bullpen noise and distractions. Depending on the type of work and the complexity of projects, either approach—privacy or large arrays of staff—can be successfully used, or even mixtures of each. The housing style should consider what type of creations are desired, while being sure to emphasize the laboratory attitude, promote teamwork, and foster sharing of expertise.

What was ideal at Hughes was its **Canoga Park** site in the San Fernando Valley, home to almost 4,000 personnel designing precision-guided weapons as part of the Missile Systems Group. The facilities and surrounding environment had the appearance of an elegant university campus. Low-rise buildings enclosed a gardened boulder-decorated pond with bubbling fountains. Surrounding terrain had open views of the Santa Susana mountain range, often used in movie thrillers of adventures in the Old West. (These were fondly called Oaters, referring to the many star horses.) This atmosphere certainly inspired creative ways of thinking each day, and motivated staff professionals to participate in educational opportunities to keep their skills honed and current.

Missile Systems Group, Canoga Park, California
(courtesy of UNLV)

Apple Home Base

The chosen site for the **Apple Campus** in Cupertino, California, in 2017, is blessed with a pleasing climate, proximity to the wonders of San Francisco, as well as the nearby educational excellence of Stanford, Santa Clara, and University of California campuses at Berkley and Davis. It provides a most invigorating and comfortable work location for inventive genius. The firm was founded as Apple Computer in 1977 by Steve Jobs and Steve Wozniak. They established a credo of intuitively forecasting what consumers would desire in the digital world, with a strong emphasis on equipment performance superiority and ease of use, coupled with artistic appearance of the product. Now named Apple Inc., it rapidly grew in success and size, becoming by 2016 the most highly valued corporation on the NASDAQ Stock Market. Citizens of all nations recognize and are eager to obtain their own iPhone, iPod, iPad, iMac, and the advanced software those contain. Another fine Apple policy is to update that software for customers free of charge over the Internet. By 2016, worldwide employees numbered 116,000.

Apple Park, the Apple headquarters in Cupertino, California, is startling in concept and size. A circular structure one mile in circumference consists of eight segments separated by open atriums. Each contains four above-ground stories and three levels beneath. All roadways and parking arrangements are down there, out of sight. Exteriors are made of glass, providing spectacular landscape views. The thirty-acre central courtyard features fruit orchards; symbolic of a vital footprint of California history, a decorative pond, as well as walkways and seating clusters for employees. Much emphasis has been placed on being environmentally green: selection of construction materials, chemical usage, large capacity renewable energy, and recycling of waste. Wow! What a magnet for stimulating creativity in employee minds!

Apple Park, Cupertino, California
(courtesy of Apple)

Shaping Production Flow

Arranging facilities with many workstations for people doing repetitive functions can be made for reasonable comfort of the employees, but is usually governed by geometric layout of assembly sequences. Physical ease is arranged with adequate spacing, ergonomic shaping of spaces and tools, environmental control, and some way to minimize noise. Continual attention must be paid to staff safety since many operations require use of hefty machinery and risky temperatures, flammable materials, and particles that could be inhaled. Effective manufacturing methods for efficient high-speed and large quantity electronics began in the late 1950s. The procedure was to divide the total processes into stages and analyze the equipment and workforce size required so that the elapsed times for each assembly step would be identical with all the others, ensuring an even and consistent flow down the line. For example, if an assembly step takes twenty minutes and the subsequent task requires only ten, arrange two feeder stations to supply step two. This also becomes the layout template for the building, with the objective of assuring minimal transportation distances between assembly stages. There is always room for creativity and innovation in arranging more efficient and better equipped manufacturing facilities.

And More Goodies

There should be adequate dining facilities for all employees. Most firms operate a cafeteria, often managed by a catering outfit, with facility maintenance performed with internal staff. Such setups provide an economical and time-efficient lunch period. It also gives a daily opportunity for employees to socially mix and exchange useful ideas, another boon for teamwork. As a special benefit for senior management, many companies have an executive dining room. This is intended as a reward for advancement, as well as a comfortable and secure meeting place for the "bosses" to exchange individual thoughts regarding corporate policy, strategy, and short-term tactics. Such topics are usually inappropriate to share with other employees, especially in the formative stages of possible changes in operating procedures. In the case of Hughes, when I arrived at the corporate executive level, I abolished that privileged luncheon arrangement. My view was that there was far more need to personally mingle on an informal basis with as many staff personnel as possible. Leaders must continually find ways to listen to the opinions and suggestions of every active participant in an enterprise while still remaining somewhat detached from individual friendships. It is mandatory to receive constructive information from anyone, but establishing close personal relations can lead to biases and unease in giving unpleasant orders. What better place to do this random and informal mixing than in a casual cafeteria? Unfortunately, some senior executives did not go along with this concept so dined out or had lunch delivered to their private offices. Many small start-up businesses today provide small kitchens adjacent to work areas and allow food preparation and consumption by staff members at any time. That surely is a casual ambience to encourage a relaxed workforce! Also, because of high-performance computers and Internet linkage, many firms allow (and encourage) some employees to perform their assignments at home.

Professional staff morale and personal health can be greatly aided by the firm providing an exercise facility. This should be available for use without charge and include adjacent shower stalls. Within corporations before 2000, staffers could appear during lunch breaks or after hours.

Nowadays, the conversion to casual working hours allows usage at the option of individual employees. The usual hardware installed would be treadmills, rowing machines, stationary bicycles, and weight-lifting devices. Providing the space and equipment can be relatively inexpensive; however, recurring operating expenses can be significant. A trained attendant should be present to minimize injury hazards, and significant insurance coverage must be maintained. When well equipped and used, such facilities will increase staff morale productivity and will likely reduce expenses on the medical benefits traditionally made part of every salary package.

Flexible Talent Clustering

Gathering people together to accomplish an objective should be set up in an **orderly but flexible manner**. As the number involved grows, more attention is needed to arrange meaningful clusters. Drawing rigid boxes on an organization chart may often be too inflexible, but there can be much confusion among individuals if there is not a clear line on decision-making, resource allocation, risk management, scheduling, and sequencing of project stages. Informal and helter-skelter arrangements will soon be fraught with chaos. A person assigned to be a leader of any unit must be given adequate authority and be held accountable to do those things and properly inspire the other people in that unit. History shows in most enterprises that unless the tasks are relatively simple and repetitive, a single individual cannot successfully lead more than ten direct reports. Following this guideline may require many levels to format a large company; this may become overly cumbersome and keep the management too distant from staff members. A remedy to avoid possible stodginess is to delegate adequate authority to as low a level as possible. This means finding leadership excellence, but it has the great benefit of

Lego, Orderly but Flexible

adding professional challenge and spice to aspiring staffers. If the mutual goals require creative thrusts into unknown regions, direct communication between all organizational levels is essential for complex information to flow freely. Often, the person lowest on the totem pole may have the best ideas to solve a group's problems. Having such ingenuity retarded by a slow approval chain can be quite inhibiting to rapid forward movement by the firm. When Hughes reached 85,000 employees, there were seven organizational tiers, but there was usually reasonable direct contact that could skip several layers when good ideas were hatched. However, it was vital that any directional changes that might emerge were passed down the chain. Orders could not be issued directly to a worker by a senior executive.

Traditional structures are categorized into line and staff functions. The term "line" means the chain of personnel that directly create, develop, test, finalize, manufacture, deliver, and support the product line. "Staff" includes legal, finance, marketing, human resources, quality standards, and technology strategy. These functions principally provide advice to senior executives and also monitor compliance with corporate policies and government regulations. They usually participate directly in customer contact, contract negotiation, centralized asset control, and long-term direction of technology research. Each staff cluster is inhabited by trained experts in these respective professional fields. These staff sizes must be well controlled since costs to maintain staffing are classed as overhead, a possible excessive burden on product pricing.

Cluster setups for manufacturing are usually arranged by grouping in work specialties and shop-floor layout. Categories include machinist, welder, casting and plating, plastic bonding, electronic assembly, material procurement, product testing, packaging, transportation, and quality control. Practitioners in these specialties can be very well trained and become experts as they gather hands-on working experience. Depending on the factory layout, large numbers of those doing these jobs can report to a single supervisor. Unfortunately, the supervisor's flexibility for location placement and job assignment may be quite limited since many employees may also be governed by a labor union. Their operating ground rules are usually set to a national industry standard, which may not even fit the work being done locally. Some Hughes electronics manufacturing plants had to negotiate with the national Carpenters Union, a residue from 1945 when the H-4 *Hercules (Spruce Goose)* was being constructed of wooden materials. How bizarre!

Another fine motivator for creative people is to allow great flexibility in working hours. Modern organizations permit staff members to form their own start and stop times each week. All that is expected is output matching the goals set for each individual. Such malleability has become easy with computer Internet linkage and data storage. Many staff jobs can be performed without face-to-face contact with teammates. Much of the work can be accomplished at the professional's home, saving valuable time that would otherwise be lost in commuting. Another way to avoid that curse is to provide home-to-work shuttle service. The vans are equipped with power and high-speed Internet connections, allowing people to work while underway. An effective morale booster can apply with those who must be in the facility workplace for equipment use or close contact with other workers.

Forty years ago, the traditional workweek was 8-hour days for five consecutive days (not long before that, it was 10-hours for six days, or even worse). Instead, all employees can enjoy many three-day weekends. This is done with five 9-hour days, followed by the "normal" two days off. The next week has only four 9-hour days, rewarded with that delightful three days on your own. Many other flexible arrangements can be made to enhance morale without loss of work output, in many cases also garnering an increase through better productivity from happier employees.

Corporations attempting to achieve great advances in complex technology development must carefully assemble the engineering and scientific staff members. Individuals may desire to focus on theory, physical research, practical design, test verification, or perhaps broad-scale exposure to all development facets. Creative people are not fond of unusual constraints or close supervision. Much flexibility in job assignments can be a key to ongoing staff morale and progress; long-term placement in a routine and repetitive position can be anathema. College majors are usually in engineering disciplines of electrical, mechanical, civil, chemical, and software, as well as the scientific leanings of mathematics, physics, and chemistry. A sensible corporate policy is to require all inexperienced college-degree hires to progress through a planned rotation program, a sequence of four 6-month positions in diverse functions. This will season and mature them, as well as yield clues about their best performance functions. All the above disciplines will be needed, but the emphasis will vary in the proportion needed for differing projects. Forecasting the mixture of disciplines needed for many future years requires much insight and wisdom. To lessen the Ouija-board risk, every organization must promulgate staff members to become multi-disciplined. This can be done by appropriate and changing job assignments to broaden their experience. A planned assignment rotation program for new employees can be extremely rewarding. Providing for continual education and a strong encouragement to participate in those are essential to cope with the rapid progress in all technical disciplines. Weekly lunchtime talks given by internal employees from diverse development projects can fertilize the imagination. A fine example of such were the Hughes Fellowship and Advanced Technology Education Programs, described in chapter 4.

Guiding Complex Projects

With the company staff being clustered by skill or by product type, it is necessary to create a centralized guiding force to coalesce the many segments of a complex product. This function is called **program management** (PM) team. That PM position is empowered with corporate-level authority over all participants in its project, regardless of their organizational location. The team's many tasks include marketing, proposal preparation, contract negotiation and recording, financial budgeting and control, master scheduling, technology integration, motivating the entire workforce, maintaining customer satisfaction, as well as reporting status and problems to senior management.

To win a new development project placed by a customer into source-competition, a contractor's effort goes through many stages and involves diversity of skills. Most of these activities also apply to thrusts into new product lines that are entirely supported by internal research and development funding. A proposal team prepares a response to the customer's objective. It contains a detailed conceptual solution, shows proof of feasibility as a master schedule, and plans for technical and schedule risk alleviation. A competitive price bid, formulated by senior management, is offered. After much discussion with the buying agency and a favorable selection, the firm negotiates detailed legal, financial, schedule, and technical requirements. Upon contract award, participating corporate organizations with needed specialty skills are assigned design and development of their segment of the system. These separate segments are directed by their own managers, but must meet the administrative and technical requirements stated by the PM. The PM provides overall program direction, issues schedules and budgets, defines the system architecture and detailed technical requirements, and closely interacts with the customer. The project proceeds through design, development, rigorous testing of prototypes, and a manufacturing trial run, followed by full-scale production and customer use. Ongoing efforts include logistical supply of repair spares, operator training, other support functions, and sustaining engineering to remedy faults and create performance improvements. Continual contact with the customer occurs throughout all stages. It is important to separate in the corporate hierarchy this central function from the performing organizations to preserve objectivity and wide vision. The number of staff people in the PM cluster depends on the size and complexity of the project. Sometimes only one professional is anointed with these responsibilities. A large and difficult project may require the PM staff to be about 10 percent of the company's total headcount assigned to that program.

The PM is blessed with having full corporate authority, matching the responsibility. Performing well requires training, seasoning, and paying attention to the lessons learned from other program successes and failures. There are many selection qualities for a PM: have adequate technical knowledge; think like a general manager; exhibit coaching leadership by persuasion and influence; be an enthusiastic advocate of the customer and the project's mission; be comfortable and adaptive to rapid change; be quickly decisive; admit errors and rectify them immediately; communicate well with individuals and in public; demonstrate good understanding of people's attitudes and needs; and always practice and advocate high integrity. This large array of traits is difficult to find embodied in individual candidates. However, starting with a good personal character as a base, many heads of organizations rapidly learn and adapt well in the high-stress job of a PM. As they grow in experience, the great ones will place difficult issues in context, describe the rationale for all their decisions, and continuously report progress to their senior executives as well as to their entire program team.

Defining What to Do Technically

The PM's technical strong arm is called **system engineering** (SE). Systems are typically made up of parts, or components, which are combined into units, which are in turn combined into subsystems, which then interconnect to make up the total system. The physical segments are controlled by a central computer's software, which is often more technically complex than any of the individual physical assemblies. This master software needs to be interactive with an operator to adapt to the particular situation. Either subcontractors or different organizations within the company can be responsible for separately developing each subsystem. System engineering's job is to ensure that the hardware and software come together as a whole and perform as the customer intends it should, including melding smoothly with other customer systems.

Visualize a three-dimensional puzzle with many small interlocking pieces that move fluidly, sing music, are of differing colors and temperatures, and are composed of different materials. All must somehow be harmonized into an operating whole to perform some exotic task. Some of the pieces will interact with others, changing size or shape to accommodate a piece that cannot change. Most people are familiar with the outcomes of these complex puzzles: harmonizing the many pieces of hardware and software in a linkup for worldwide air traffic control; a surveillance and navigation system aboard ship; a dynamic tool for managing a business conglomerate's finances; and a nationwide communication satellite network. Numerous separate devices must rapidly perform complex functions that join together to carry out the total mission.

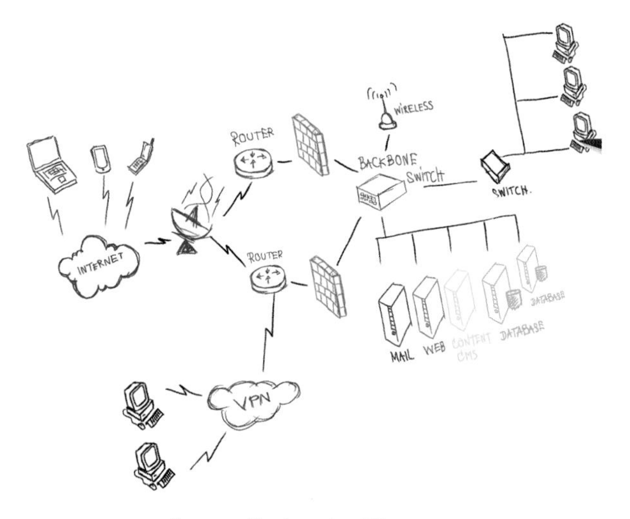

Systems Engineering Diagram

System engineering is the center for technical decision-making within the PM's team. Many tasks begin after a thorough understanding of the overall requirements: devising the system's architecture; analytically setting overall and component performance and reliability goals; defining interfaces and interactions among subsystems and with other systems within the overall project; determining whether to develop a component in-house or subcontract it; specifying limits of size, weight, power, and cooling; negotiating with the system's carrier vehicle maker or user's site to accommodate the equipment's needs; constraining the hardware and software configuration to maximize repairability; prescribing manufacturing and field maintenance standards; setting reliability and repair goals; and specifying the testing and evaluation of fully integrated prototypes. System engineers continually negotiate and compromise with the detail designers, making many adjustments in order that the sum of all the parts achieve the contract's overall performance level. All these tough tasks must be carried out within the cost constraints of available funding budgets. This practice, called "making tradeoffs," occurs when a better-performing device could offset a sister segment experiencing a shortfall or when an initial design approach proved unsatisfactory. Ideal qualifications for staff personnel doing this work are broad technical vision, intuitive perception of potential risk areas, and a persuasive and tactful skill to give directions to performers in other organizations.

4

PHILOSOPHY, COACHING, AND REWARDING

DURING AND AFTER paradise and structure have been established, there must be careful thought to form the management style to be used. Each firm has a different purpose, a way to get things done, and faces a vast array of external constraints with which to cope. All staff must be inculcated with understanding and believing in the company's goals, funding limitations, and nominal method of operation. That staff must feel that they are contributing to worthy ends, are recognized for what they accomplish, and are adequately compensated and rewarded. A happy set of employees will boost productivity, assure healthy recruiting, and earn the positive recognition of the firm by customers and the public.

Here it is appropriate to recommend six books that describe how innovation has been nurtured in recent history. These can help form a firm foundation of understanding how the creative process works successfully (see appendix for details):

The Gene: An Intimate History by Siddhartha Mukherjee
How We Got to Here by Steven Johnson
The Innovators: The Discoveries, Inventors, and Breakthroughs of Our Time by John Diebold
The Innovators: How a Group of Hackers, Geniuses, and Geeks Created the Digital Revolution by Walter Isaacson
The Technologists by Matthew Pearl
Hughes After Howard by D. Kenneth Richardson gives details of the striking ingenuity that sparked Hughes Aircraft Company to earn recognition as the world's leading outfit for the invention, development, and manufacturing of military electronics.

Philosophy Umbrella

Corporate leaders must establish an operational aura that will continually inspire the staff to pursue excellence. A unifying family team spirit can greatly motivate daily activities and mollify concerns when something goes wrong. Employees always feel good knowing they are doing something worthwhile. It can be especially magic if their team is rated to be the finest in that field of endeavor. Being the best is an excellent attraction for new talent. The philosophy and operating standards emanating from top leaders also provide a protective umbrella over their workplace. Although high-quality performance will be expected, there will be reasonable margins of judgment to accommodate occasional shortfalls. Below are three corporations that set high standards for the entire workforce. More details are given for Hughes because of the author's personal experience.

The Iconic Bell

Bell Laboratories, headquartered in Murray Hill, New Jersey, was the sine-qua-non for facilities, talent, organization structure and operating policies. It accomplished numerous significant technology breakthroughs. Bell Labs practices and style have been emulated by many other dynamic and forward-looking companies, such as Apple, Hewlett Packard, Hughes Aircraft, Intel, Lockheed Skunk Works, Texas Instruments, 3M, and Xerox.

Established by Alexander Graham Bell in 1881, this unit was initially named Volta Laboratory and Bureau. As the American Telephone and Telegraph Company (AT&T) rapidly expanded, this segment was named Bell Laboratories in 1925. Its purpose was for general research, plus an additional goal of devising and designing operating hardware for communication networks. That equipment would be manufactured by its corporate sibling, Western Electric Company.

The parent company's performance excellence plus business and political savvy propelled AT&T to become a monopoly controlling the operation and advanced development of all U.S. long- and short-distance telephone services. The resulting massive inflow of revenue enabled

an outstanding corporate strategy to plow a large proportion of earnings into internal research and development (R&D). This rich supply meant that Bell Labs never felt impeded by resource limitations. Such a boon fostered what is essential in research efforts: sustain long-range goals. Many new ideas take a long time to mature into moneymakers. Unfortunately, those research glory days were considerably curtailed in 1982 after an eight-year legal battle. With the admirable objective of restoring competition in this important national service, the United States negotiated a breakup of AT&T into independent regional business entities. The parent company then only retained control of the long-distance net. This was a good change in order to reestablish that eminent credo that "competition is best" (see chapter 9), but was devastating to the supply of huge research funding for Bell Labs. Since then, the iconic outfit has greatly withered and is now owned by Nokia, a successful high-technology outfit in Finland.

In its heyday, Bell created the transistor, charge-coupled electronic devices, the maser, information theory, software programming languages, and numerous telephonic improvements. In basic science, it achieved understanding of the fundamental structures of magnetic materials and glass, the wave theory of matter, high-resolution fluorescence, microwave radiation in the Universe, and details of quantum mechanics. For these things, Bell proudly sported eight Nobel Prize awardees. This peek at history clearly illuminates the wisdom, vision, and few limits on free thinking that were espoused by the leadership at that famous laboratory. How sad that many possible new discoveries in such wide-ranging regions of knowledge may be severely retarded by scarce funding support.

3M Shines

A noteworthy innovator with a significant corporate credo to stimulate creativity is the **3M Company**. Founded in 1902 as the Minnesota Mining and Manufacturing Company, its objective was to produce and market rock materials for grinding wheels. Business declines caused management to search for and develop other product lines. A big success was marketing many applications of adhesive materials. The company has grown to over 80,00 employees located in thirty-two nations, with sales forces in two hundred countries.

To assure a continual flow of new ideas, 3M supports thirty-five laboratories worldwide. Familiar consumer brand names include *Scotch Tape, Scotchgard,* and *Post-it,* all part of a broad product diversity array of 55,000 varieties. In addition to public consumers, beneficiaries of their useful outputs are firms in manufacturing, packaging, cleaning, medicine, and entertainment. Many of their creations occurred after a laboratory surprise or an error led to a new compound that started another line of sales. A fascinating example is how Patsy Sherman evolved Scotchgard, as described in chapter 5.

Author John Diebold, in his book *The Innovators: The Discoveries, Inventions, and Breakthroughs of Our Time,* clearly describes a corporate philosophy that has resulted in this plethora of unique products in widespread areas of commerce, based upon the belief that everyone, in differing degrees, has an innate urge to create something never done before. We will urge them to find ways to turn those ideas into marketable products. There are five major tenets:

1. Set a difficult challenge: each organizational unit's goal is that 25 percent of annual sales must be derived from things developed in the previous five years.
2. Create a sense of responsibility; each unit operates and is held accountable as its own independent business entity.
3. Make available adequate resources to support new thrusts: product income credit belongs to each unit, and reasonable portions may be allocated to research and development. Often, those may be supplemented from other sources. In all cases, technology skills and understanding within all of 3M belongs to all and will be shared through cross-pollination.
4. Assure sponsorship to individual thrusts: supervisors are to encourage and support promising new ideas hatched within their staff.
5. Provide appropriate rewards: title recognition for an achiever's success is done with dual ladder provisions. Those with leadership qualities progress up the management tiers; those with superb R&D talents are publicly recognized with increasingly weighty titles. These parallel chains have equal prestige and compensation.

The Hughes Style

Following are six guiding philosophical tenets that helped **Hughes Aircraft Company** attain its acclaim in the world of advanced electronics. Primary to setting the proper corporate aura is to promulgate a strong and meaningful mission for the efforts underway. This goal must be clearly understood by all, be a unifying force to muster team spirit, and perhaps inspire an emotional drive within individuals. Some examples to be considered today are such goals as to find a remedy for the most aggressive types of cancer; cure blindness; improve the quality of education; maximize safety of self-driven automobiles; find a way to gather and dispose of billions of tons of nondegradable plastic now dispersed throughout the world; perfect ways to use coal-based energy without pollution; and, most difficult, find an acceptable way to limit human population growth (in only ninety years, it skyrocketed from 2 to 7½ billion, and is currently increasing by 1½ percent annually!).

Motivation of everyone at Hughes was strongly fired-up by the perceived threat posed by the Soviet Union. Its stated objective was to attain international political and economic dominance. The U.S.S.R.'s vast technology investments stirred a competitive drive for us to create superior operating performance of all types of weaponry. National survival was at stake for forty-five years. That, indeed, was a meaningful emotional driver. With such a compelling long-term purpose, teamwork and sharing of expertise were self-generated and sustained, requiring little management push or concern. To keep the goals alive, the staff should be made aware of in-house achievements as well as the comparative progress of the opposition. These information sources were often hampered by military security limits, but any publicized accomplishments made by the Soviets usually sufficed to stir our always-do-better spirits. Similar comparisons with breakthroughs done by our domestic arch competitors ignited an even more intense bonfire of eagerness to excel.

Hughes staff members were also imbued with the mystique demonstrated by the company founder, Howard Hughes. He was judged by the public as eccentric in personal behavior, castigated by our revered Dr. Simon Ramo as technically incompetent, found to be an inept manager by his customers, and thought to be a poor aviator by professional pilots. Nonetheless, Mr. Hughes had an internal motivation of great importance to creative engineers in our business: always design things far superior to what others can do. Another component was to be secretive about all actions. This offers a protection from competitors harvesting inventions, minimizes public awareness of actual activities underway, and provides a real boon to protecting military security. That magic mystique lent a unique aura to the Hughes organization as viewed by others. His other fanatical driving force was to perform with excellence and create things that perform better than those of anyone else. Anything less was totally unacceptable.

There must be adequate funding and a long-term viewpoint practiced by the corporate management. Most new development requires significant design effort and intensive evaluation before being considered ready to manufacture. There can be many surprises in attainable technology, uncertainties in schedule, and alterations of customer desires. There may even be the need for starting an alternate design should the primary approach fall short. To assure success and avoid staff panic, there must

Howard Hughes in His Office
(courtesy of UNLV)

be a calm and steady hand at the wheel. Long-term strategies can be very difficult. Since Americans are impatient by nature, stockholders can demand quick successes on a quarterly basis, or competitors may be first to the marketplace. However, it is vital to provide stability and tolerate unsuccessful design attempts. In the case of financial strategy, investing a significant portion of earnings on a regular and steady basis into research and development must be done. This was relatively easy at Hughes while it was privately owned. A dominant corporate goal was growth in product lines, expanding market share, and increasing total sales. These emphases were far healthier for technical advances and staff morale than focusing on increasing profit percentage and milking cash away from investing in the future.

Any company's inherent value will increase with an emphasis on recruiting superlative talent. This may result in higher salary expenses, but far more inventive minds and higher productivity will yield more creative products, improve competitive status, and buoy internal satisfaction and pride, providing an incredible pool of brainpower to apply to any new design challenge. Individual performance levels ranged through good, very good, brilliant, and extraordinary. Some of the latter could dream up an entirely new concept that would yield a unique and world-class product. These Renaissance men and women could not only visualize the end result but also predict development complexities and the likely problems that would have to be solved. Their almost ethereal insights in any topic were a beacon of inspiration for others. Sources we successfully harvested for new bright staffers were the top graduates of engineering schools, university professors looking for better careers, and other electronics firms that did not have stimulating professional assignments or were cursed by poor recognition and compensatory structures. The magnets drawing fine talents to us included that Hughes mystique, a chance to create at the technology forefront, being part of a united family team, ideal working environments, great salary structure, and continual opportunities to be educated about rapid advances in technology. Very attractive to top college graduates were the work/study Fellowship Programs, adding one hundred new Masters candidates per year, equally split between UCLA and USC. We also sponsored two Doctoral Fellows per year at Caltech. The company promulgated an active, ongoing internal updating service for employees called ATEP (Advanced Technology Educational Program). Expert insiders and notable outsiders, such as Dr. Richard Feynman, passed their wisdom in classes to any interested staffers. Over a twenty-year period, ATEP gave 1,780 classes with over 63,000 enrollees.

Another corporate philosophy was to encourage a dynamic team spirit among all employees: we share a major goal and can best succeed

in that mission by cooperating together as a family. Simple ways to pass along this concept are through statements such as, "let us unite to meet our common foe" and "join together to save our bacon." This approach enabled a bountiful sharing of knowledge. When a problem arose, contact between managers in remote sites could find an expert experienced in that issue. That staffer would be loaned for a reasonable time to the area of need. This allowed quick resolutions of road-blocking surprises in development, testing, manufacturing, or field use. "Family" meant professional sharing, not lovey-dovey social bonding. There still was plenty of room for employee competition for acclaim and reward. Those solving someone else's dilemma gained great and lasting recognition. Mentors are positively regarded by everyone.

The overriding company credo was to operate with integrity in business and technology development. That word "integrity" is fraught with dramatic differences in interpretation, based on moral standards, business practices in other nations, and competitor skullduggery. Our corporation placed high value on education, but as President Theodore Roosevelt once said:

> To educate a man in mind and not in morals
> is to educate a menace to society.

What do you do if the only way to effect a sale to a country will require payment of a bribe to the procurement officer? The best way is to make the offer far better in hardware performance than any other option that nation can find. Surely that procurement chief will set aside the shady money temptation. But suppose your competitor finds a covert way to deliver the bribe and wins the program before you can act? Ouch! And how to win against an overseas supplier whose costs are subsidized by its government, making for a lower offering price? Hopefully, the same hardware performance margin will prevail. (An example of "offer the bestest and the mostest" can be found in chapter 7.)

The difficulty of "correct" analysis of technical and business integrity is the Fukushima Daiichi Nuclear Power Plant 2011 disaster. The nuclear facility was disabled and partially destroyed by a very strong earthquake followed by a major tsunami. The location had been selected by the Japanese government. Professional advisors and a construction budget had been set. With concurrence by all involved, structural requirements were set to assure survival in an 8.0 Richter scale shake, a magnitude higher than any recorded in the previous four hundred years. Few tsunamis of the size of this one had ever occurred at that site. But the massive tremor experienced was 8.3, more than twice as violent as that 8.0 limit. To build structures to survive that giant level would have been far beyond foreseeable technology limits and would have been completely unaffordable for the nation. So, was anyone at fault regarding the ethics or integrity of the decision-makers to implement this important power plant? It seems very unlikely. Operating any business always involves setting limits due to affordability. Integrity results from revealing properly and in a timely manner the likely outcome and potential risks of the project. And, of course, when you try anything, be well informed, forecast the risks, and do not violate any laws of your region.

Here is an endorsement of the philosophy practiced at Hughes Aircraft Company. Gloria Wilson was at the forefront of advanced radar design and was superb in many technical and leadership roles:

> The dedication of the people I worked with was extraordinary. The Hughes culture was to design and build the best for our military. People's lives depended upon what we built, and we had that foremost in mind as we worked in teams to accomplish that goal.
>
> The company made a strong commitment to education, supplementing college courses with the Advanced Technical Education Program operating on site. These courses kept our engineers on top of our game in all technical subjects. We were taught by visiting professors and by senior members of our own staff. This dedication to ensuring technical excellence by Hughes is a significant factor in our success over the years. It also engendered a cohesive team spirit.
>
> The cooperation between all parts of the company was unique. Specialty roadblocks could quickly be remedied by calling an expert anywhere within our large geographically dispersed staff. We had a true family constantly surrounding us, sharing knowledge and inspiring success.

Choosing Great Coaches

When an organization is being formed, and ever thereafter, coaches must be found and properly placed in the structure. The term "coach" is used herein to properly describe a leader of a creative technical cluster, where maximum delegation should be prevalent. This is to avoid negative connotations that may cling to such terms including manager, supervisor, chief, boss, and master. (I once enjoyed being favorably called "the big kahuna," a Hawaiian prophet.) Specific behavior and skills needed as a supervisor can vary widely, depending on what needs to be accomplished. The desired characteristics can be quite different than those observed during that individual's hands-on job assignments. Those who select individuals for promotion must have good judgment and intuition. True, most effective abilities can continually improve through acquiring experience, training, mentoring, and a careful planning for ramping up the novice. Exposure to increasing stress and uncertainty should be synchronized with the coach's growth in demonstrated capability. Even the choice of military leaders may be difficult, although the unit's goals may be relatively simple (move forward, not back). The obey-my-orders doctrine can be significantly enhanced when the leader exhibits persuasiveness and exudes an aura of well-deserved authority.

Sensible guidelines for the selection process start with assuring the candidate has an educational or experience foundation for **understanding the product line.** If the projects are financial, pick a person rooted in comprehending economics; grocery stores need commodity and distribution background; those in department stores surely need expertise in clothing and the dynamic vagaries of women's fashions. If the product has high or exotic technology, it is essential that the entire leadership ladder is populated by those with intensive education in engineering or science. Too many technology firms have gone on the rocks when the top-line appointees have superlative financial or legal backgrounds. Such professionals, although very competent, have been chosen by the corporate board in order to maximize stock market value. Since they do not understand the product complexities, they can lose sight of the length of time needed to mature a product for the market as well as the need for continual design improvement. As Aristotle said, about 360 BCE (nothing has changed since then!):

Union Cavalry Officer
(iStock/Woodkern)

You must first learn to row before you try to steer.

The percentage of earnings committed to IR&D usually atrophies markedly. Senior executives with technical roots may be somewhat inept in the essential financial control procedures, so it is vital that they be advised by extremely competent financial and legal staffs. This suggestion is based on personal observations by the author. It is quite likely that other firms using different guidelines for leader backgrounds have been quite successful. Some individuals can be superstars with superlative judgment on all business and technical challenges, but those are rare. Mr. Hughes, regardless of his technical abilities, at least set high standards and made demands for hardware performance excellence, was able to properly observe results, suggest improvements, and provide almost unlimited funding to reach those lofty goals.

The second clue to observe is **people skills**. Few things can be achieved these days without the enthusiasm and cooperative spirit of a team. Its head coach must know how to inspire, ignite enthusiasm, comfortably control or re-steer effort, and set attainable goals. It is optimum for any person in these positions to be able to converse and exchange ideas with anyone, regardless of their training or working level. Instilling comfort and trust must always be done. Good ideas can spring from any source, as will be discussed in chapter 5. People skills may be inherent, but can also reach excellence with continual practice and mentoring.

The third personal characteristic to note is the candidate's ability to **get things done**. This skill is usually inherent in an individual, but can improve with time. Self-confidence will naturally grow as successes add up and failures have been overcome. There is much to be said for persistence, determination, and zeal to achieve. These traits in a coach will greatly and positively influence the other team members. Some highly self-confident individuals are great with talk, but weak in accomplishment. Others demonstrate little humility, which quickly erodes everyone's feelings of trust. A prime illustration of "having the right stuff" is Dr. Art Chester, the outstanding leader (and a true Renaissance man) of the diverse HRL Laboratories (formerly Hughes Research Laboratories) in Malibu, California. He was a fine and respected coach, with ultimate humility. He once said:

I may know something about most things
three inches below the entire ocean surface,
but don't ask me for any details deeper than that.

The fourth selection check is whether the candidate **proceeds with integrity** in the profession. Integrity in their social life should be their private concern, but the company reputation may also be at stake if poor actions become known. Such a person should be aware of their risk of discharge if cultural moral codes have been transgressed or if the public has become aware of strange activities. Integrity within the firm is also extremely important at all times. It is astounding how one small step in the wrong direction can adversely impact customer confidence, stock market value, employee stability and trust, and the possibility of costly legal actions. The old phrase that casting a small stone can cause a giant avalanche certainly applies here. Integrity in an individual is usually obvious to most observers. A questionable attitude and behavior are usually easy to spot. If not noticed by senior executives during the coach-selection process, then lack of integrity in their actions must engender quick remedial action.

Hughes Coaching Style

In chapter 3, a number of books are cited describing successful ways to organize and control businesses. Each provides great advice backed by historic successes. When fully absorbed, they reveal how extremely varied and complex it can be to choose the best path. The "best" methods must account for the business type, competitive concerns, backgrounds of the employees, and the ambient culture. Hughes evolved a series of optimal behavioral characteristics of those who coached creative elements of the staff. Here is a brief synopsis of what worked favorably. Significantly, these credos helped us to encourage free-thinking and to delegate maximum authority and responsibility.

There are several no-nos. Coaches in the work environment must guard against becoming **pedantic, dogmatic, and dictatorial**. Of course, when emergencies arise, such forceful methods must prevail. Ways of being at the helm must be well accepted by those pulling the rowboat's oars. Direct criticism or second-guessing, especially about work in progress, is most unwise; such can be stifling to an individual's attempts to try new things. The coach must **know and understand the staff** assigned. Team members will vary considerably in abilities, methodology, and productivity. Some demonstrate small contributions on a regular basis; others may hatch brilliant ideas only sporadically. As for personalities, here's an interesting quote of Kingman Brewster Jr.:

> You and I know that there is a correlation between the creative and the screwball. So we must suffer the screwball gladly.

Blending a cooperative team of diversity can be a challenge. A good solution is to imbue everyone with the belief that we are all together in a great cause and will share the prestige and rewards for success.

The coach must properly execute tasks toward the **long-term goals**. Primary here is to **inspire** dedication, persistence, and excellence. Decisions must be crisply done. Much uncertainty reins in development projects. Outcomes from some moves may be unpredictable. One useful concept from the wisdom of baseball star Yogi Berra is:

If there is a fork in the road, take it.

I used this, tempered with the confidence that choice was correct if analytic data indicated 80 percent positive. If new data shows up, the direction can be changed. Most of what the staff had done in that interval would be of value; far better than frittering their time away as the coach did not choose a course and cope with uncertainty. Naturally, the coach has to be straightforward and take responsibility for the actions. Always listen to suggestions and absorb new information as the project encounters reality. Another challenge is to decide when what has been done is good enough, so the staff can apply its efforts elsewhere. The judgment must come from what is in the customer agreements and whether improvements can be made with little further effort.

How does one best steer the boat? The word "inspire" has a magic ring, and doing this requires a lot of practice. Earlier mentioned was the need for people skills. Additionally, acceptance by the team depends on whether the coach sees the big picture, has reasonable intuition, believes in what is being done, and can speak with authority while still being humble. That is quite a mix of assets. The staff also abhors close supervision or imposing arbitrary procedures. Mark Twain said:

It's good sportsmanship to not pick up lost
golf balls while they are still rolling.

Give the staff time to correct their own errors. Know what they do, but don't closely monitor how they do it. Sometimes humor can soften the stress of giving commands or negotiating agreements. It was most amusing to observe Walt Maguire, PM of the F-14 *Tomcat* fire control system and the *Phoenix* missile, maneuvering the hardware design leaders to accept a challenge. There had been a serious test failure of a prototype in flight, and a possible fix was contemplated. The designers strongly stated, "We cannot perform this redesign within a year, and it will cost more that $1 million."

Walt quietly replied, "OK, why don't you show you cannot do it in six months for a half-million?" Outraged, the designers departed the meeting, but sure enough, they successfully matured the fix in five months and spent only $400,000.

Ongoing mentoring markedly matures, strengthens, improves the abilities, and boosts the morale of the workforce. Wisdom and long experience must be passed on to others, whether in functional disciplines or for grooming new coaches. It can be done formally in individual or group meetings, but is usually most effective when done informally. This author had five excellent mentors who contributed immensely to effectiveness in every job assignment. One memorable happening was with our company president John Richardson (no relation to me, but I would have liked to have some of his genes) about 1980. He and I had just concluded a difficult meeting with several senior Navy (USN) officers who were discontented with our progress on two major programs. As we strolled to our car, I thanked him for the many bits of advice he had given me for years; I then inquired, "What else should I focus on?"

His immediate response was, "You do all the big things just right; we only have a few little ones left." What a superb vote of confidence!

Recognition and Rewards

It is most reprehensible when someone grabs credit away from the person who really made some magnificent achievement come to pass. Such often occurs in the political milieu, but in the disciplines of technology, creativity, and business operations, this must be strictly verboten. When a new discovery or procedural improvement idea is successfully hatched, appropriate **recognition** can come in several forms. The coaching staff must always be aware of what is going on in their area and make sure that public knowledge of the individual or team is well known. Enthusiastic awe by peers provides an enormous prestige boost. The coach can also award positive salary adjustments, organization title boosts, and assistance in publishing papers or attaining patents. Such measurable notes of appreciation will spur others to strive for their own breakthroughs and stimulate repeats from those lauded. The inverse, where the leader takes credit, will obviously brutally demotivate the true performers. So, another watchword for senior executives is to observe such evil and effect quick cures.

The **salary structure** is strongly affected by competitor firms seeking new staff, availability of skills needed, recognition for individual contributions, and amount of responsibility assigned. There are also constraints imposed by the government and by labor unions. To maximize employee joy, it would be most ideal if your firm was viewed as the land of milk and honey, but that is not possible because of all those factors. The company's existence as a business may hinge on total cost of operations.

Keeping the entire workforce reasonably content, yet still filled with zeal to excel, is crucially affected by the fairness and possible growth in personnel compensation. Salary levels and employee benefit programs are the keys to positive staff morale and attracting new talent. There are many complex and interactive considerations that must be co-mingled for long-term company stability and well-being. Most staff members compare local practice with those of other firms, as well as somehow comprehending what their office peers make (which should be kept private by each individual). There will be unrest if large discrepancies persist. Sometimes they will be discontented when the company rewards an individual who demonstrated great productivity or conceived a unique new procedure or product. There often are political attempts to mandate equal pay for all people in similar work assignments. Though such would satisfy the we-all-should-be-nice approach, it is completely counter to our cultural drive for surmounting odds and receiving due rewards. History shows that positive incentives breed great advances for the entire society. Keeping things equitable can be greatly aided by always arranging semiannual interactive discussions between individuals and their coach. This exchange of views should be frank and open. Misunderstandings and suggested improvements should be most welcome by all parties.

Recently there has been much more discussion and dispute about equal pay for equal work, particularly as that applies to ethnic and gender differentials. In 2017, the *Economist Magazine* revealed that the Women's Policy Research organization heralded a 21 percent gender pay gap in 2015. The article then described a more recent detailed analysis performed by Korn Ferry Hay Group. That study was far more accurate since, in addition to compensation level, it included the essential factors of function being done and the nature of the business. The data base was enormous: 20 million employees in 25,000 organizations in 110 countries. With all three factors involved, the analysis showed a gap only 1.6 percent. The major pay differentials were due to "contribution," or company benefit garnered from the worker. Four stages of contribution were cited: (1) person works with help of others; (2) tasks are performed independently; (3) output is accomplished

Efforts Rewarded

through others—this individual does effective coaching; and (4) this person leads the firm's strategy. Most women were clustered in stages 1 and 2. That shortfall can be remedied by placing more women in leadership positions. Another notable study highlighted that in many cases when such had been done, those firms improved their profitability by 15 percent.

In addition to vacation privileges, staff **benefits** usually include some medical insurance, life insurance, sick-leave provisions, and retirement packages. The Hughes arrangement of the latter was to give the employee the option of withholding 4 percent of their salary plus another 2 percent more from company coffers. These sequestered funds were managed by a corporate center; we were the nation's only large firm to invest those so well that the total value far exceeded the accrued future payment liabilities. All of these benefits were costly and were part of inter-company competitive tactics. The ongoing and future costs of benefit programs must be carefully monitored and closely controlled. In the last few years, employee benefit programs were carelessly expanded to keep governmental staffs happy, resulting in bankruptcy declarations in Iceland, Ireland, Greece, and Puerto Rico. It seems quite foolhardy to allow retirement after only twenty years of service and guarantee payment of 80 percent of the final salary level for the rest of that employee's life. Many governments make long-term obligations well beyond their future ability to pay. There are no investments or reserves set aside to assure solvency when the recipients become entitled to receive funds. Once that pledge has been made, it sets an ongoing precedent. It will be politically almost impossible to adjust the amount downward to fit future budget limits. That bugaboo may also apply to private corporations.

Commuting Gem

Cessna Citation II

Those who ascended the corporate ladder (both in management and in scientific tiers) at Hughes could gain other good things. Travelers in any level of responsibility could be **transported by air** on company business by a Cessna *Citation II* aircraft. Two of these, each with a capacity of eight passengers, were housed at Van Nuys Airport in the San

Fernando Valley. Daily flights from and to Tucson, Arizona, were regularly scheduled. Other on-demand trips were to many company plants in California, Colorado, New Mexico, and Texas. More distant sites usually required commercial service because of the Cessna's flight range limitations. Middle-level managers also could obtain a **new automobile** for commuting from home. These were provided every other year, after which the employee could acquire it at the extant used-car price. After we became a General Motors (GM) subsidiary, any employee could purchase any current model for a significantly discounted price.

Another treasured benefit was to attend a management **offsite conference** with your spouse. These three-day meetings were scheduled annually at comfy resorts in Santa Barbara, Palm Springs, or Rancho Bernardo. These were not only stress relievers for the participants, but also greatly enhanced our family aura and global understanding of the entire corporation's strategies and problems. Another productive practice that can help bonding middle management teams is to arrange periodic informal afternoon parties for spouses. These can be held in classy hotels or country clubs. An overview of what the organization is doing and plans for the future are summarized. A caring spouse can greatly mellow an employee's stress by comprehending the achievements and risks being experienced in the workplace.

Executive Extras

Senior executives in large corporations merit special consideration in compensation. They are responsible and accountable for the operational stability and future survival of valuable assets and the employment of thousands of workers. In recent years, excesses have become rampant; many executives are paid far beyond their contribution to their firm or to society. This author believes that some strong measures must be taken by stockholders and boards of trustees. Personal behavior and practices of corporate leaders vary widely, depending on their personalities and upbringing. Other factors are strong social pressures, practices in similar industries, evolution of cultural standards, and board of trustees vigilance.

Practices did change significantly during my forty-year career at Hughes. Many of the following odd anecdotes could have been standard practices in the aerospace industry as a means of rewarding excellent performance by leaders facing complex responsibilities. Most were abolished and likely are no longer practiced. Some changes followed harsh chastisement of the individuals. Public exposure has imposed some substantial restrictions on what should be allowed. Here are some "specials" paid for by the company; hopefully, most have rightfully disappeared: construct a barn on a private ranch; install home improvement alterations; anodize a forty-foot sailboat mast; in the company plant prepare a movie film of personal World War II fighter combat history; and use company aircraft for flights to resorts in the United States and Mexico. (With retrospective regret, I did enjoy several of the latter). Other lesser evils: always use first-class air flights; set no limit on elegant hotels or dining or boozing on business trips; have regular landscape services at your residence, and have details of personal finances and checkbooks performed by the staff.

Appropriate **performance rewards** were salary, title prestige, achievement recognition, annual bonuses, and stock options. The latter usually consist of a set amount to be purchased by the recipient within three years at the same price as when awarded. If that stock declines in price, no benefit to the awardee will result, so the option will not be exercised. However, if the market price has risen, there will be a gain on those shares purchased, but they cannot be sold for the next six months. This is an excellent motivator to get the executive to do their best to improve company results. However, the practice has been heavily criticized: some unethical executives fixed the financial books for those three-and-half years, then cashed in and departed. When the truth was revealed, the stock price for everyone else sharply declined. For many years, such options were not available to Hughes staff members since the private owner, Mr. Hughes, did not permit any equity sharing. This changed in 1985 when we became a General Motors subsidiary; its public stock certificates could be offered as options. Prior to then, that lack of equity participation provided some rationale for the excessive perks criticized in the previous paragraph.

Another fine perk is the award of a **country club membership**, with expenses paid. This was a real privilege, and also provided classy venues to entertain customers and trade ideas with other business executives. It was an honor to join the Bel Air Country Club, seeing many celebrities and viewing from the fairway the façade of the elegant residence of Howard Hughes during the 1940s. Additionally, the top four corporate chiefs also could use a **limousine service** for their daily commute and in-town trips. Some employees felt this was frivolous and overly expensive. However, there were several productivity gains garnered. In addition to personnel security

(senior leaders can be a target), there is relaxing convenience, short times between distant meetings, schedule flexibility, and the opportunity to work while in transit.

As President, I did much valuable work in my one-hour each way journey while being driven by a well-qualified driver. This included reading paperwork and effecting many long-distance telephone calls to Washington and overseas customers. A real time saver was to use one of two Bell Ranger helicopters to make frequent zippy trips to many of our plants in thirty-five California cities. This avoided likely ghastly jams in the Los Angeles freeway network. Longer distance journeys could be aboard our Grumman *Gulfstream II*, capable of one-stop trips to our beloved Pentagon. It was a boon to bypass the commercial flight arrangements, but I always felt that this expensive vehicle was a bit cheeky for me to use.

Limousine Service

Beating the Freeways

Executive Flight

5

INVOLVING THE ENTIRE COMPANY

As ILLUSTRATED PREVIOUSLY, new concepts can spring up from many sources. As stated in chapter 4, it is optimum for any person in a position of influence to be able to converse and exchange ideas with anyone, regardless of their training or working level. Instilling comfort and trust is always wise. Many times, those doing the hands-on tasks are best equipped to visualize improvements in process or entrenched habits or conceive unique approaches to attain a goal. It is most valuable to learn how to understand and converse in the varied dialects of people with different backgrounds. These exchanges must be done with comfort and sincerity, and not as an assumed posture. The coach must also maintain a leadership stature, without becoming socially attached by casual friendship with staff members. Such model people-skills sometimes may be inherent, but individuals can also reach excellence through continual practice and mentoring.

Hierarchy Fences

It is difficult to understand why substantial technology advances and human lifestyle benefits have been made in the cultures of India, Great Britain, and even in the United States. Social hierarchies would seem restrictive for free exchanges of ideas. Subservient individuals will be greatly inhibited from suggesting new concepts that may be viewed as radical. The upper level tiers, both economically and socially, also will assume that the "lesser" folk are inherently incapable of independent thought or of generating breakthroughs.

Very dramatic and publicly well-known is the traditional **caste system** practiced in India. There are four tiers in the totem pole, with status based on birth heritage, education, and type of work performed. Atop the crown are priests, called the Brahmins; followed by the Kshatriyas (warriors); then the Vaishyas (skilled trades). Servants and subordinates in the bottom layer are termed Sudras. Beneath those are the Chalandalas, who are not even considered to be part of the caste structure. Doing the most menial work, the "untouchables" literally may not be physically touched by caste members for fear of permanent contamination. Certainly, never ask one for fresh ideas. Perhaps this arranged social ranking might perpetuate merit based on inherited traits, but it is ruefully self-sustaining. There is little upward progression allowed regardless of brainpower. Once born-in, one stays in and has little chance to speak above the imprinted status.

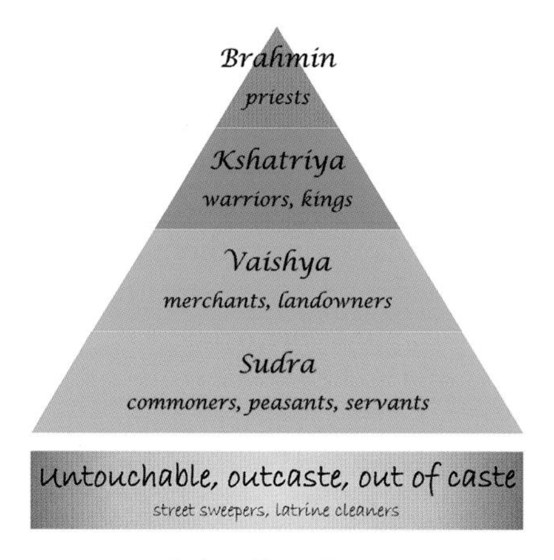

Indian Caste System

Similarly, England is still working to remedy the illusion that **as-born-status** is more important than character or ability. The British hierarchy since 2000 has undergone much improvement regarding the social barrier limits. There is far greater interchange than in previous eras. However, the traditional regal and peerage inheritance practices are still in place. If you are born in a titled family, you will inherit a fancy title, regardless of ability. (It surely must be nice to be respected and honored as part of the privileged aristocracy.) As in all nations, one's educational history, rightfully so, is held in high esteem. If you carry that magic Oxford or Cambridge banner, many doors will open.

British Naval History Postage Stamps

At least the official classification structure in the United Kingdom (The National Statistics Socio-economic Classification, or NS-SEC) is based upon achievement and level of responsibility. One can see from the descriptors of each level, that upward movement can be based on talent and achievement, not just by a title that was passed from an ancient ancestor.

The eight levels range from Group 1 = professionals and senior managers; through Group 5 = lower supervisory and technical; down to Group 8 = long-term unemployed. We have all advanced a long way from the Middle Ages. That amazing array of eleven rigorous class barriers were emperor, king, prince, duke, marquis, earl, viscount, baron, baronet, knight, esquire, and gentleman. All others were classed as serfs or slaves. Few free interchanges of ideas were tolerated. How in the world did any change in traditional human behavior happen?

National Statistics Socio-Economic Classification, United Kingdom

1	**Higher managerial and professional occupations**	**11.1%***
	1.1 Large employers and higher managerial occupations	(4.3%)
	1.2 Higher professional occupations	(6.8%)
2	**Lower managerial and professional occupations**	**23.5%**
3	**Intermediate occupations**	**14.0%**
4	**Small employers and own account workers**	**9.9%**
5	**Lower supervisory and technical occupations**	**9.8%**
6	**Semi-routine occupations**	**18.6%**
7	**Routine occupations**	**12.7%**
8	**Never worked and long-term unemployed**	—

* *Data* : Those currently employed Labour Force Survey Winter Quarter 1996/97 (excluding Northern Ireland). N=63,233 (may not add to 100% due to rounding).

Even in the United States there remain many social practices that hamper open discussion. All are being worked on by the general society, but progress has been slow. These barriers may stem from differences in cultural or racial practices, gender, age, speaking ability or accent, and even the state or county where one grew up. There are also rankings done by profession, management titles, name of university attended, religious leanings, wealth, and political beliefs. Fortunately, the relaxed U.S. informal social behavior places few inhibitions on speaking out, when appropriate, directly to the person most likely to take action. At Hughes, there often were meaningful discussions jumping between any organizational tiers that enabled great concepts to be acted upon. In order to save time and avoid frivolous distractions, it was essential that the idea espoused had been well thought out and the employee was well prepared. These discussions were done while still preserving the authority needed at each management position. There were few instances where a top executive, upon hearing a new concept from a subordinate-tier individual, would issue action orders without going through the proper management chain.

A shining example of our informality occurred during the Paris Air Show in 1985. Dr. Allen Puckett, waiting alone in the company's marketing chalet, took the opportunity to try out a new Japanese computer-printer. A French Air Force wing commander entered the room unannounced and, not realizing who Allen was, casually asked about how the printer worked. Dr. Puckett cordially described all the technical details he had so far observed. The wing commander was later flabbergasted upon hearing that he had been chatting with the CEO of the gigantic and notable Hughes outfit. He remarked that nothing like this could have happened in contacting a French company officer: rigid hierarchy, cultivated caste system, and formal protocols would never allow a senior manager to mingle with any unknowns.

The preceding statements about India, Great Britain, and the United States may appear to have a political flavor. It certainly would be most unproductive to favor or emulate the Socialist or Communist advertised credos. History has clearly shown that national systems adopting these have failed economically. It is an illusion to believe everyone is the same, and all natural resources should be shared equally and managed by a central agency. Several things are missing: recognition and reward for achievements, meaningful competition, little incentive to improve personal life, collective rather than individual thinking, and no self-reliance. These factors are deadly to inspiring creativity and ingenuity. Perhaps the theoretical lack of hierarchical separation can allow free exchange of inventive ideas, but that hope soon disappears in countries that have tried this political slant. It seems that the U.S. approach has performed optimally, in spite of some grievous excesses in business practices. Our culture usually is successful in properly differentiating people based on their inherent ability, educational level, demonstrated productivity, and zeal to perform positively. The theme championed herein is to honor and reward talent as well as to remove artificial communication barriers imposed by birth-heritage, political leanings, or organization rank.

Here's a good spot to pass on the wisdom of leadership expert Gordon Tredgold, warning of what he calls the "seven deadly sins of leadership." He provides seven no-nos to avoid: (1) becoming too distant; (2) being too friendly; (3) feeling a need to provide all the answers; (4) acting too hands-on; (5) taking too much credit; (6) using control rather than influence; and (7) continually micromanaging. Operating without these flaws will greatly nurture and accelerate the fruitful personal interchange of valuable thoughts.

Ingenuity Blossoms

Many people are wary of technology because of its cost, the possibility of evil results, or that there are few things left to invent. As far back as 1834, and again in 1902, the newspapers misquoted the Patent Department's incumbent chief as recommending to the President that the department be abolished, saying "everything useful has already been invented." Fortunately, such nonsense did not prevail, and the federal agency still thrives. There is a tangible offset to the erroneous perception that everything has already been done: since 1960, there have been over 55,000 patents issued for diverse uses of the laser, as mentioned in chapter 11.

Adding to the gems of invention described previously, what follows are several more vignettes of creations that came by surprise to this professional. These all further illustrate that anyone with resources or an entrepreneurial bent must always stay tuned and be alert to spot breakthrough concepts.

Road construction, beginning in 1820, used compacted crushed stone, invented by Scottish engineer John McAdam, but often became rutted and dispersed. In 1902, English surveyor Edgar Holey observed attempts to remove an oil-tar spill on a macadam roadway using slag from a furnace. The oil and slag actually bonded into the crushed rock, thereby greatly improving the smoothness and ruggedness. His bright observation resulted in *tarmac*, still used today throughout the world as a paving material.

Tarmac Road

Teflon Pan

Dr. Roy Plunkett at Chelous Jackson Laboratories in New Jersey patented and brand-named *Teflon* in 1945. He had discovered that compressing tetrafluoroethylene caused it to polymerize and become a solid waxy material. This process revolutionized the plastics world and continues to be used prolifically in aerospace, architecture, manufacturing processes, electronics, and communications.

Swiss engineer George de Mestral, on a hunting trip during World War II, became curious about why cockleburs instantly attached to his trousers and to his dog's fur. Study under a microscope revealed that inset hooks in the thistle crown stuck to the natural loops fabric and in canine hair. He and family members in the weaving business found a way to manufacture strips of these using mohair. *Velcro* flexible and easily released fasteners are now found worldwide in numerous applications.

Velcro

Nachos

Here's ingenuity that sprang from an urgent need to avoid calamity to the business. Each Friday for several years, six women from Calexico had strolled across the border to Mexicali for a casual luncheon. Their favorite rendezvous was a small restaurant run by a charming and enthusiastic chef. One week, they arrived early and the kitchen larder was almost empty. In desperation, this clever Mexicano heaped cheese atop piles of tortilla chips, placed those briefly in an oven, and with a flourish, presented this unique new dish to his favorite clients. They were very delighted, and soon spread the word to their friends at home. Thus emerged an international favorite: *nachos*.

An accident culminated in a popular new product after 3M's Patsy Sherman spilled a chemical mix on her sneakers. Very annoyed when trying to clean them, that spot retained the original color, while the surrounding area was faded or permanently discolored. Although she was a skilled and well-trained chemist, she also certainly had an enterprising instinct. She visualized that this fluid compound could become a large-scale new product to act as a protective surface over any type of fabric. Thus, *Scotchgard* was born in 1952.

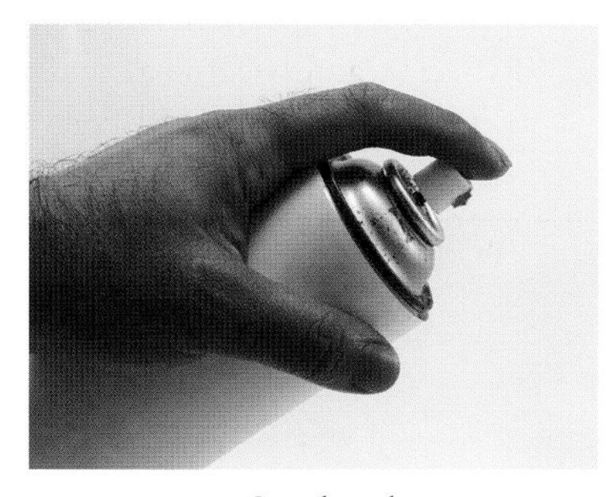

Scotchgard

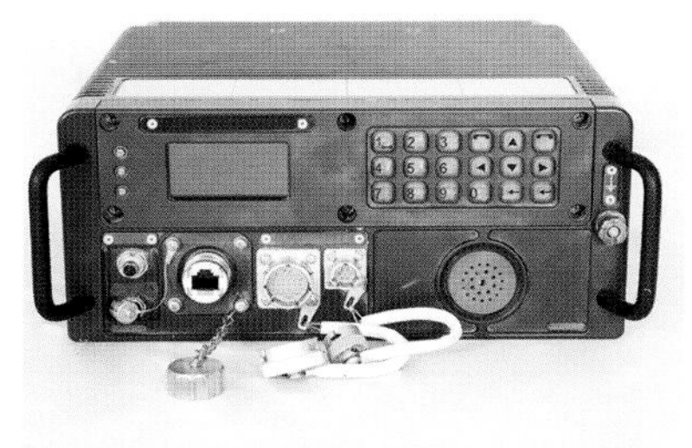

Frequency-hopping

An astounding fresh technology idea sprang from an unexpected source in 1940. Famed movie star Hedy Lamarr had technology interests in addition to her acting profession. She did a bit of hardware tinkering, and she even interested Howard Hughes into offering free support from his company's engineering staff and facilities for some experiments. Her whiz-bang concept was a way to avoid German countermeasure signals from foiling radio-guidance of Allied torpedoes. The new technique was to rapidly change the frequencies of the guidance control signals being sent by the submarine. That architecture could also be used to send secret radio messages that could not be intercepted.

Since Ms. Lamarr was musically adept, she chose the number of shift positions to be the same as the familiar eighty-eight piano keys. The frequency switching was timed by using a player-piano activation scroll; the sender and desired receiver controlled these tapes. The sequence of frequency shifts was done randomly for each individual mission. She teamed with musician George Antheil to prototype her concept. Now called *frequency-hopping*, the concept was patented in 1942, but not formally adopted by the U.S. military until 1962. Apparently, there were some closed minds in some sectors of the Pentagon during World War II.

This architecture is now in common use for all forms of communication.

Marketing professionals often come up with spectacular extravaganzas that fulfill challenging goals. Tom Shaver, global merchandising manager at General Motors, brainstormed something that was likely to be placed in the Guinness Book of World Records. Hearing in 1985 that GM was about to acquire Hughes Aircraft Company as a subsidiary, he hatched a marvelous way to welcome those 85,000 employees and have them enjoy being part of their new family. Of course, this event would also include a strong marketing twist. He observed:

> I knew Hughes employees owned many cars that were not GM vehicles. Since Hughes was based in the Los Angeles area, I had the idea to rent Disneyland Magic Kingdom, place GM products throughout the park, and open the park exclusively to Hughes employees and their families.

Disneyland, Anaheim
(iStock.com/smckenzie)

Disneyland in Anaheim was rented for this purpose for nine consecutive nights. Each Hughes employee was given up to seven *free admission* tickets, covering all rides, entertainment events, food, and drinks. Circling overhead was a blimp with flashing lights displaying, "Welcome to the GM family." This gala happening was thought to deserve that coveted Guinness award as the largest private party held over successive nights in a single location, with 180,000 attendees. The new subsidiary members certainly were most pleased to be welcomed in such a showy celebration. One could hope that many GM car sales resulted.

Leaders of any enterprise may benefit from this Walt Whitman quote:

Now, voyager, sail thou forth to seek and find
(to which should be added, "and communicate well to others.")

Teaming Design and Manufacturing

For most firms producing tangible products, manufacturing is the key to business and financial success: it's the best source of earnings, customer satisfaction, and enhancement of one's reputation for future sales. In designing its production facilities, the company must think through many important considerations to ensure an outfit that would yield low-cost and high-reliability devices on a predetermined schedule.

Achieving this can be particularly difficult when those devices are very complex, require extreme precision fabrication, and might never have previously been made anywhere. Such high-technology manufacturing particularly needs willing cooperation between design engineers and production personnel. Open ears and a willingness to adopt ideas from persons with differing skills can overcome many obstacles to effecting a producible product. Manufacturing organizations should have considerable management authority placed by corporate executives, and those corporate leaders must pressure the engineering leaders to place high priority on solving design-related difficulties discovered on the factory floor. Delays in smooth production flow can be very costly. The inherent laboratory-like practices may still remain: creative engineers doing their initial design may have little interest in, or knowledge of, the most efficient ways to ensure easy, reliable, repeatable, and low-cost manufacturing. Cooperative efforts yield greater results than when organizations work independently.

The best way to effectively transition from something defined on paper to the reality in production is to incorporate a preproduction practice stage in the program master schedule. One example of such a transition at Hughes was that of a new infrared sensor system, a very complex mechanical, optical, electronic creation using exotic materials and requiring tricky fabrication and assembly techniques. For a year, overlapping the final prototype proof testing, a large team comprised of equal numbers of design and manufacturing experts pored over the configuration details. They were empowered to alter designs slightly, invent new production processes, carefully outline operator instructional documents, define the timing sequence for information transfer, and conceive short-cut corrections if things went wrong on the floor. Test runs of fabrication and assembly processes refined their reliability and repeatability.

As a result, full production experienced little difficulty, met schedules, and even achieved lower cost than anticipated. That team effort proved a worthwhile investment. It avoided what often happens in similar situations: antagonistic confrontation between the design and production camps. Even with a referee, such behavior breeds festering long-term ill will, time delays, higher expenses, and even foments pervasive internal political warfare to establish a winner. The better teaming method, with its mutual understanding and communication methods, also helps the manufacturing staff respond rapidly and flexibly to the inevitable flood of future design changes that always occur in the birthing of leading-edge devices. Changes arise from incorporating remedies to problems discovered in further testing, reliability or production improvements, field-use experience, and customer-desired performance improvements.

Electronics Assembly Lines
(courtesy of UNLV)

Once the product design is understood, a crucial manufacturing staff decision is whether individual parts or subassemblies should be made internally or bought from a qualified supplier. This decision hinges on the internal fabrication capability, the suppliers' reputations (they may already be the experts in making a particular component), the quantities and delivery rates, the recurring costs, the reliability, and the customer's preference for how much of the total work should be subcontracted. Overlying all these is to fulfill the critical importance of each part in meeting the final product requirements and accounting for the expected duration of need for the particular component (it might soon be eliminated by a probable redesign). Deciding those factors demands close and continuous communication with the design staff. It is vital to establish and nourish effective, cooperative relations with many firms in the supplier base. When involved early in the design, subcontractors and suppliers can recommend alterations to match their capabilities or to lower cost. In the subsequent purchasing phase, such relationships can result in many cost and quality upgrades as well as generating positive motivation for the supplier to meet schedule and quality commitments. Whenever possible, it is most desirable to attain lowest unit price and assure positive supplier relationships to contract for a multiyear purchase. The supplier can reduce many operating expenses through bulk purchasing, fabricating at efficient rates, and enjoying stability in employment and overhead control.

Assembly Expert

Another illustration of Hughes' design and manufacturing teamwork was finding a way to efficiently and correctly make the intricate center panel for a planar array antenna used in the F-14 *Tomcat*, F-15 *Eagle*, and F/A-18 *Hornet* radars. A perfectly flat, 36-inch diameter aluminum sheet, less than ½-inch thick, had to be pierced with hundreds of cavities interconnected by channels cut into the surface. In turn, the cavities, ½- by ¼-inch rectangles, had to be precisely shaped with absolutely square corners and placed with accuracies of one-thousandth of an inch. The outer veneer faceplate for the array also needed hundreds of precise rectangular openings harmonized in position to the inner panel cavities.

The shape and sizes had to match exactly the electromagnetic wavelength of the radar's operating frequency. After a year of struggle (and quite a few animated disputes), the manufacturing staff, working with the design team, created a unique new milling machine controlled by a computer to do this tough job in a low-cost, repetitive manner with accuracy better than the levels specified for product acceptance.

Quality Turmoil

Smooth, reliable, and low-cost manufacturing requires controllable processes, repetitive task sequencing, adherence to personnel safety, and government-imposed standards. All process steps must be clearly defined and stated in terms understandable to the trained operators. Results of each procedure must be measured, recorded, and analyzed to ensure adequate yields. In the 1950s, the highest production rates at Hughes were to be found in missile manufacturing in Tucson, Arizona. The large production facility, jointly established with the Air Force (USAF), had been sited there to be many miles from the coastline to render potential Soviet bomber strikes less likely. In its first few years, it successfully delivered over 50,000 *Falcon* air-to-air guided missiles. For each production stage, yield charts were updated and displayed throughout the working floor. These provided employee awareness; with such public visibility, individuals and teammates were highly motivated to improve their performance and to recommend changes to the task procedures being used. Follow-up actions to remedy flaws were well documented, studied, and used to alter the low-yield step. Much attention was paid to quality control for all processes. The high motivation of missile employees was much enriched

Precision Machining Tool

by their awareness of how significant the products were in fulfilling a meaningful national need.

The most costly business event in Hughes' seventy-year history was what became known as the **Missile Quality Problem**, which flared up in 1984. The Tucson facility in Arizona was actively producing the antitank *TOW*, air-to-ground *Maverick* and *Walleye*, surface-to-air *Roland*, and air-to-air *Phoenix* missiles, plus ARBS bombing sights, and several other high-performance products. The most costly quality-control activity was the possible rejection and discard of large quantities of "one-shot" devices, such as the tracking xenon lamp to be installed in the *TOW* missile's tail for guidance tracking. Fifty of each group of 5,000 were tested; if more than one did not operate, the entire lot was discarded.

Tucson Facility, 1951
(courtesy of UNLV)

Compliance with voluminous customer-imposed standards, specifications, and contract terms needed continuous attention and care. In spite of careful monitoring of all production processes, Tucson got in big trouble on the little things in the 1984 quality assessment fiasco. In all other Hughes manufacturing facilities (and certainly in Tucson after the 1984 difficulties), long-term, high-volume, and excellent product performance earned high marks from all customers.

The 1,000-pound AIM-54 *Phoenix* missile, prime weapon for the Grumman F-14 *Tomcat*, was an essential element of the Navy's Fleet Air Defense network, slated to prohibit bomber and cruise missile attacks against large sea battle groups. The weapon could engage targets one hundred miles distant, and six of them could be simultaneously in radar guidance against six widespread hostiles. Mr. William "Will" Willoughby, a dedicated civilian bureaucrat in Washington, D.C., was head of quality for all U.S. Navy (USN) equipment procurements. His push for quality improvement was most appropriate. Although we were meeting all terms of the existing contract, he announced that until Hughes improved its manufacturing quality, the USN would no longer accept *Phoenix* missile deliveries. We did not sense that any contract adjustment of our fixed price would be allowed, even though many changes to operations and equipment might be needed to satisfy Mr. Willoughby.

Supreme quality is always a meaningful and high-priority objective. However, perfection can be costly: infinite quality usually can come only after infinite expense. Missiles are important weapons, so product quality must not be treated casually or unduly compromised, and there must be a continuous effort to control and improve manufacturing processes, both in-house and by component suppliers. The defining element is the contract between the customer and the producer, which dictates the level of performance demanded for the end-product's price. The Tucson plant had continuously met or exceeded all *Phoenix* contract terms and conditions. Flight test results were far better than stipulated in the contract, with a much higher hardware testing success rate than had been experienced by any other guided missile manufacturer in the world.

The USAF, which was in charge of quality monitoring at the plant, and the USA were both very satisfied with the Tucson products delivered, but they had to join a Department of Defense (DoD) alliance with the USN. The government confrontation with Hughes was difficult to accommodate since the company was exceeding contract requirements. We were most willing to implement changes to improve quality, but we were unwilling to bear the added implementation costs or to be denied delivery acceptance. It was always possible for an inspector to find minute and meaningless hardware deficiencies.

Analysis of the plethora of government specifications applicable to *Phoenix* for materials, processes, and tests revealed 675,000 pages of documented requirements! Total "compliance" was impossible, especially since many items specified in one page conflicted with requirements in other documents, and some statements were obsolete. It would seem very reasonable to judge total quality based on end-item reliability and performance accuracy rather than on the little itty-bitties.

As President of the Missile Systems Group, I made a very regrettable mistake by fostering in the company an overly defensive attitude. The worst blunder was accepting an interview with Ralph Vartabedian, an excellent aerospace reporter for the *Los Angeles Times*. During the interview, I described our manufacturing practice: if an error is found in the hardware it is fixed before the next manufacturing step, and if that happens again, either the process or the design is changed. One example I showed Ralph was a small scratch in an aluminum frame that had a preservative

Soldering to Etched Circuit Board

coating added to shield it from any lifetime environmental harm. This repair should have been acceptable, but the government inspector judged "insufficient manufacturing quality" and said that the entire expensive subassembly should be discarded. What should be paramount to the customer is that the total product meets its contract requirements. The interview was a lively one, and unfortunately, the next day the *Los Angeles Times* featured a passing remark I had made: "After all, it's just going to be blown up when it's used." That foolish statement infuriated authorities in Washington, D.C., and I think it triggered the big firestorm that lasted for more than a month. Government officials apparently thought I was too blasé and had assigned a low priority to superb quality. That perception was certainly and definitely most untrue, but I soon felt a bit paranoid and unloved!

With concurrence from our CEO, I shut down production of everything and assigned the entire Tucson staff of 8,500 people to identify and remedy any processing flaws that workers could find. It began with pep rallies for everyone. With energetic music in the background, I stated that we were in this together and everyone would be fully paid as we solved the crisis (a risky promise since with no products delivered, we had no income cash-flow, and I had no idea how long a full solution would take nor its cost). I further pledged that all innovative ideas would be eagerly welcomed and no suggestions would be ignored.

A good example of what should have been done early on in our assembly planning was pointed out by several on-the-floor staffers. If supplier components are stored in the open air for weeks, the metal connecting leads could be coated with light film. When a solder joint was completed in its installation, the joint could visually appear OK; but when undergoing vibration in the field, the joint may separate (called a "cold-solder" joint). The remedy was to short-term store before use or dip the leads in solvent before soldering. Easy fix!

Blaine Shull, an expressive and persuasive executive with immense people skills, was chosen as the new manager for Tucson operations. His personal-relations talent became vital to quick progress as he reassured both the distressed employees and the unhappy customers.

The shutdown caused great dismay in the USA and USAF with delays in deliveries of vital weapons. And the shutdown cost Hughes an unrecoverable $300 million. The crisis was resolved by dedicated actions of our staff, especially the hands-on workers, and had superb long-term results, including greatly increased customer satisfaction. After three weeks, we were able to restart deliveries. To announce the good news, I gave more pep rallies and thanked the individuals and teams for achieving our success. This time, the background music was Willie Nelson singing "On the Road Again." Many outstanding improvements resulted in every product. Quality and reliability greatly increased, and manufacturing costs were substantially reduced. If only we had had the gumption to do all this without being spurred by such a ghastly crisis.

These events were actually therapeutic throughout all segments of the 85,000-employee company. All manufacturing personnel on the factory floor did an excellent job of perfecting processes and fabrication procedures to significantly improve end-product quality. In the 17,000 staff Ground Systems Group, one team so excelled in producing high-quality assemblies that the boss of the government inspection squad reassigned his people elsewhere in order to keep them busy—the inspectors continually had found not a single manufacturing discrepancy to report. This astonishing decision so excited our on-the-spot Hughes team that they made large logo buttons to clip on their shirts proclaiming: "We Don't Need You." This provocative display horrified the company's marketing personnel, who feared customer retribution. They finally had the buttons eliminated, but our quality stayed top-notch.

Superlative Outcome

The *Phoenix* became quite successful in deployments by the USN and by Iran. Over 6,200 were produced. It was de-activated in 2004, after thirty-two years of service. Even though results of this 1984 factory-wide crisis solution were superb, my lingering personal irritation came from our being forced to greatly exceed contract requirements at our own expense; the government zealots mandated changes without spending a nickel from DoD coffers. In retrospect, if I had taken a more positive, cooperative, and enthusiastic approach, it is likely that much of our large unplanned expenditure would have been reimbursed. This is a good lesson for managers in all fields of endeavor to learn! Don't be overly defensive, and do negotiate your way through difficulties with the customer. Hughes ultimately benefited from having far better manufacturing capabilities, but there is no evidence that we gained a competitive edge. This well-publicized crisis motivated others in government agencies to start establishing second sources for procuring many products (described in chapter 9.) It is usually wise and correct to assure competition, but for items in low-rate production, the potential savings are overwhelmed by the costs of implementing the second manufacturer, a cost born by DoD, not that new contractor.

It can be declared that our Missile Quality Problem engendered substantial long-term benefits to the United States. Often great things can emerge from controversy by encouraging all individuals within a large organization to continually improve the manner and style of their tasks. Their leaders must inspire a noble team mission and actively engage the ideas of everyone who will create the final product.

Phoenix Launched from an F-14 *Tomcat*
(courtesy of the National Naval Air Museum)

6

KNOW THY CUSTOMER

UNDERSTANDING CUSTOMERS IN the commercial markets is relatively easy since your staff will probably be enjoying the same culture as the many customers in the same nation. They will speak the same language, perhaps with regional twists. Two things must be at the forefront for strategic moves: (1) earning respect for product quality and (2) maintaining continual two-way communication. Honoring and acting upon customer reactions to the product will yield handsome rewards. Some of your staff, hopefully, will have the intuitive skills to forecast achievable devices and service concepts that will attract others to new product lines. Attractive to the enormous individual consumer market will be items that improve personal comfort, save personal time, or increase entertainment value. In preparing commercial items to sell to other companies, the attractions are things that will improve their operating efficiency and quality, and perhaps enhance the appearance of their product lines. Of course, for all markets, offering a juicy bargain in price will be most enticing.

Commercial Gold

Hughes Aircraft tried a number of **commercial products**, not all of which were successful. Attempted were the digital watch, described in chapter 11; laser scanners for pricing items at grocery checkout; three-dimensional (3-D) sound system using only two speakers; computer-controlled machines for laser-cutting fabrics into patterns; high-definition TV display screens; as well as flight training simulators. Our defeats were not because we did not understand or were unacceptable by customers, but because we quickly were underpriced by other firms with low-cost manufacturing and widespread marketing expertise.

The successes we had included data-stream multiplexing for single wire linkage for multiple display outlets, night vision for law enforcement and fire-fighting, traveling wave tubes for radars and communication networks, sensitive and high resolution infrared detectors and arrays, and laser devices of many varieties. For several years, we also were in the enviable position of being the source of 60 percent of the world's satellites. Customers of those were governments and corporations providing commercial communication services. A crowning glory in the consumer marketplace was *DirecTV*, described in chapter 11. The beauty of this program was receiving continual monthly payments directly from millions of consumers, with no fee payments to intermediaries. In all these products, superior performance captured the buyers.

Seeking Military Rapport

Cultivating customer relations and earning their confidence is essential for continuing success in any business. When the product is destined for the **military**, direct involvement by observing and performing live-operating combat equipment leads to firsthand comprehension of battle events, adds reality to design decisions, and enhances open and friendly communications (one learns to speak their special lingo). Many of these experiences also bolster the ability of company team coaches to inspire thousands of employees to commit themselves to invent, develop, and produce excellence—and perhaps to visualize themselves in combat. For years, the author placed much personal attention to this vital objective, interacting with all branches of the U.S. military service. I was somewhat embarrassed in these encounters that I had not personally served in the military. During the Korean conflict, I was deferred from being drafted due to engineering studies in college. After my graduation, I passed the physical exam. Amusingly, at its conclusion, a burly Marine sergeant asked me to list in order of preference which services I would like to be assigned. My response, "Navy, Air Force, Army, Marines." He quickly said "good, we will assign you to the Marine Corps"! Fortunately, that Asian conflict was over so I was not called up. To this day, I regret that I did not acquire the training rigor and spirit of the Corps. However, during my professional career, particular emphasis was placed on air combat, matching several product lines I worked on, and fulfilling a yearning to fly agile aircraft, ingrained at the age of five.

Here are a few glimpses at the lingo varieties one can notice in the field. Command in stressful or perhaps in all situations must be clearly heard, understood exactly, and obeyed immediately. The words must be succinct, spoken in a voice with authority and clarity. Orders are given in two steps: a word to "prepare," followed by a silent space, then an emphasized word to "execute." An illustration is "order" . . . "**arms**!"

Words familiar to most people are those used to clearly communicate individual alphabetic letters such as easy, foxtrot, hotel, juliet, kilo, lima, sierra, and zulu. Air words include affirmative, base leg, Mach, 6 o'clock, slipstream, sortie, VOR, and yo-yo. On the ground you may hear declination, defilade, fall back, flank, hedgehog, redoubt, round, and shoot. At sea, words used include abeam, athwart, cat, companionway, coming, conning, hatch, larboard, and salvo. The list of unusual curt words may virtually be endless.

Over forty years, military customer viewpoints were garnered from visits to diverse areas of active military operations: fifteen Naval Air Stations, eleven different aircraft carriers, twenty-two Marine bases, twenty-five Air Force sites, and twenty-nine Army centers. All these encounters and numerous other contacts contributed to a deep and meaningful understanding of our military customers. Marketing, developing, testing, manufacturing, operator and maintenance training, and long-term product support were more correctly done because of the accumulated knowledge gained. I did my best to pass along the information and customer feelings to all employees, and I encouraged them to also understand, or possibly witness, operations in the field.

Professional practices and attitudes may differ between each of the services. What follows is based on personal observations. It is somewhat colored by the particular projects that Hughes supplied to each. Let's sequence from USMC to USAF, then to USA, and finally to USN.

Semper Fi

Products for the **U.S. Marine Corps** (USMC) included fighter radars, anti-tank weapons, air-to-air and air-to-ground missiles, bomb delivery navigation for strike aircraft, and radar warning of incoming projectile to steer return fire, tank fire control, laser target illuminators, and trooper night vision scopes. Most of these items had been initially developed and manufactured under contracts to the USN or USA. However, it was important to comprehend the differing mission executions done uniquely by the Corps. Of the USMC bases I visited, the highlights were Camp Pendleton and El Toro Air Station, both in Southern California.

El Toro provided a rich opportunity to meet and discuss tactics and training of fighter squadrons. These pilots were able to operate aboard aircraft carriers but would normally need to function effectively from crude landing surfaces in combat zones. Air intercept skill was practiced, but much emphasis was on strike support of troops on the ground in forward areas. At the time, the principal aircraft in use were the F-4 *Phantom* and the A-4 *Skyhawk.*

A special visit to Camp Pendleton provided a double thrill. A practice **amphibious assault** on the beach was a sharp reminder of World War II landings in the Pacific islands, but now employing very modern equipment. Landing craft were faster and more able to climb ashore, naval artillery was very accurate, and air cover by helicopters and strike fighters seemed extremely effective. In 1992, I toured Guadalcanal, and visualized what a blessing today's hardware would have been to our Marines fifty years earlier. Witnessing those amphibious adventures, followed by dining with the commanding officer in a tent above the beach, was most inspiring. While we munched bologna sandwiches made with white bread, he explained all the planning difficulties and continual coordination needed to successfully carry off this complex mission. Details of logistics, allowance for equipment failures, as well as synchronized timing of sea, land, and air elements were exceedingly intense. It is a wonder that all this can ever be done while being exposed to a competent hostile.

Another significant learning experience happened far inland at Pendleton to observe **practice shots** of the Hughes anti-tank *TOW*

USMC Amphibious Assault
(USMC photograph by Cpl. Jacob A. Farbo)

missile. The weapon launcher was atop an armored personnel carrier, with a moving target ½-mile distant. One attempt was very scary: when the gunner squeezed the trigger, nothing happened. This was a "hang-fire," apparently caused by a mechanical failure to start the rocket motor. Personal safety was at stake, since it was not known if the motor would ignite in attempts to physically dismantle the test. A very cautious withdrawal by the gunner, followed by appropriate deactivation procedures, resulted in no injuries.

TOW Gunner
(USMC photograph by Cpl. Chris Lyttle)

Blue Skies

Our company had been involved with the **U.S. Air Force** (USAF) and its predecessor since before World War II. We became mutually dependent for air intercept weaponry in 1946 and served them well for decades. There were numerous projects with advanced performance in ground, air, and space missions. In forty-seven years, Hughes developed and manufactured for the USAF equipment for thirty different types of combat aircraft, sixteen guided missile varieties, and several highly classified communication and surveillance satellites, as well as sophisticated ground-based command-and-control networks. Throughout those many years there was a continual strong bond.

Here is the view of USAF Gen. Richard Myers, former chairman of the Joint Chiefs of Staff, the highest U.S. military position:

> I remember Hughes being very customer oriented, with great people. These people not only had great technical skills, but also had a good grasp of how the equipment they were developing would be used by the men and women of our military. I always came away from discussions with Hughes staff believing they understood our needs and had great technical capability to produce what we needed.

The author's exposure to the USAF way of doing things revealed two very different styles: (1) those on flight duty and (2) the administrative legions. Experiences with the latter were on a daily basis, since that branch of the service was responsible for the oversight of all DoD work being performed at Hughes. As a result, I was always wary during any contacts with them since their duty was to find noncompliance, error, or legal misconduct. Those are most appropriate, but some reporting was of minor incidentals, so it appeared that motivation of some individuals was always to find something to criticize, whether worthy or not. That seemed to be a real drag to forward progress, like slogging through a dismal swamp. Of course, there were notable exceptions of responsible USAF staffers who wanted to help row the boat toward our mutual objectives. All these relations, both pro and con, were cast in sharp relief during the dark days of the missile manufacturing turmoil described in chapter 5.

Seeing the "active operating" characters was positively inspiring. They seemed to be driven to perform their important missions with vigor, excellence, and the will to outperform any potential hostile. The three notable bases visited were Wright-Patterson Air Force Base (AFB) east of Dayton, Ohio; Eglin AFB southwest of Valparaiso, Florida; and Nellis AFB in the Las Vegas Valley, Nevada.

Wright-Patterson is the center for managing all development programs, and we were in constant contact, having at least thirty active projects under USAF contracts. Its location was placed in the town where the Wright brothers had their bicycle shop and fashioned the first heavier-than-air powered flying machine. That miracle had first flown successfully in December 1903 at Kitty Hawk, North Carolina. How the world has changed since then!

Nellis AFB for many years was the controller of Area 51, also called Dreamland. That site was used for testing and evaluation of advanced devices being developed for the CIA and DoD. All activities bore the rigor of high national security protection. It was exhilarating to be there to witness combat in AIMVAL/ACEVAL and **Operation Red Flag**. These projects yielded progressive training for fighter pilots. This was done over large areas of desert and mountain terrain. Mission exposure to hostiles was most realistic, using accurately simulated combat encounters. Actual Soviet aircraft and ground-based tracking and missile control radars were employed. Further reality was to pit all types of U.S. aircraft against varieties of Soviet fighters to develop countering tactics and equipment improvements. Dozens of flying machines could be simultaneously involved in these dynamic encounters. Sophisticated U.S. ground-based radars were also used to verify reflected return signatures of all aircraft types.

Red Flag Encounter
(USAF photograph by Tech. Sgt. Larry E. Reid Jr.)

Aircraft and pilots from many other allied nations also participated in these tense melees. Israeli pilots were adept in dogfight scraps with Egyptian fighters flown by novice pilots. Victories were virtually assured with cannon or machine-gun fire, or even by forcing the opponent to maneuver into the ground. Their confident comments were: "Who needs a guided missile? All you need is a good airplane and a well-trained fighter pilot." This, of course, was nonsense in the bigger picture. (In Desert Storm conflict, twenty-eight shoot-downs of Iraqi aircraft were done at long range using Hughes *AMRAAM* missiles.) Interactions between wide varieties of fighters using Russian tactics and well-trained pilots gave far different results. One clever French unit, operating F-5 *Freedom Fighters* to simulate a Russian squadron, realized that their aircraft were not equipped with radar detection equipment. They purchased some automobile speed-trap "fuzz busters" from an auto shop, mounted one on top of each fighter's control console, and successfully detected incoming radar illumination, which they used for effective evasive maneuvers. As stated in many of these chapters, ingenuity can blossom anywhere.

Eglin AFB is both an operational aircraft site and the principal facility for evaluation of new equipment. On one occasion I was privileged to present the annual **Hughes Trophy** to the 58th Tactical Fighter Wing. Every year since 1953, this prestigious award was given to that year's best USAF fighter unit. This time the winner was an F-15 *Eagle* squadron. It had achieved sixteen Iraqi aircraft kills in Desert Storm, the highest score of any USAF unit. It was a real honor to meet and shake hands with these scrappy pilots and to receive a framed poster with their signatures. At the formal dining celebration that night, my wife was seated next to Gen. Charles Horner, who had commanded Allied Air Force operations in that conflict. He introduced himself to her saying, "I'm General Horner—just call me 'Chucky.'" That's a great way to foster relaxed conversation, and is typical of the active flying folk in the military. The trophy is on display in the USAF National Museum at Wright-Patterson AFB.

Hughes Trophy
(USAF photograph)

Several of our USAF programs were performed as subcontracts to other aerospace firms, notably McDonnell Douglas and Northrop Grumman. Good relations were vital. Fortunately, there was little discomfort with the manner of conversation and discussion of objectives. We all were in the same profession and had similar academic backgrounds. Here is an example of how well things went: Dr. John F. Cashen, former Vice President, Advanced Projects and Chief Scientist, of Northrop Grumman stated:

Hundreds of dedicated Hughes employees pioneered a new chapter in airborne surveillance. The team at Northrop Advanced Projects found both the outstanding cooperation and the challenge of managing such technical excellence to be exhilarating. The program created many lifelong friends within and bridging the two companies. In 1980, Hughes competed with Westinghouse for the B-2 stealth bomber radar. Both companies had outstanding radar divisions in those days, and it was a very close competition. Both proposed radars would have met all the specified requirements. I was part of the selection board at Northrop's B-2 Division, and my recollection is that the deciding factor was Hughes's unique design, although it was not the most advanced. Contrary to my doubts, the design proved highly reliable, and it gave Hughes the winning edge with an antenna whose weight was significantly less than the competitor's. The B-2 became operational with that design, and for twenty years performed with distinction. The collegial attitude and technical excellence of the senior Hughes antenna engineers, along with the confidence demonstrated by Northrop and the USAF, led to an exceedingly successful program outcome, greatly enabling America's premier strike capability, the B-2 *Spirit* bomber.

Ground Control

The types of equipment for the **U.S. Army** (USA) included fire control systems in four helicopters, two aircraft, and two tanks, five guided missiles, two artillery counter-fire systems, two trooper night vision aides, three laser-spot projectors, four secure radio systems and six air defense networks. The most important bases I visited were Fort Bliss in El Paso, Texas; Redstone Arsenal adjacent to Huntsville, Alabama; and Fort Sill north of Lawton, Oklahoma.

Fort Sill, established in 1869, is the home base for artillery training. It was here that Apache chief Geronimo was held captive in 1886. A continuing Native American legend tells that he had once evaded capture here by leaping with his horse from Medicine Bluff into the nearby river! It was a thrill to see the historic parts of this famous fort. Fort Sill gives civilians an opportunity annually to watch **mobile artillery battery** emplacements in the field, observe fire zones, and witness live shots of an M712 *Copperhead* projectile. This is a cannon round guided to its target by its internal seeker tracking a laser spot projected by a forward trooper. The significant finale performance is quite emotional. Accompanied by bugle calls and drums, horse-drawn howitzer units appear from behind a ridge; guns are placed, hand loaded, and fired. Suddenly, it is 1863 at Gettysburg: the equipment and uniforms are recreations of those of a Union Army artillery. That day provided an astonishing demonstration of the sharp contrast between inaccurate mass-fire barrages of the old days and the precise guided projectiles today.

1863 Revisited
(iStock/MindStorm-Inc)

Redstone Arsenal is the site of the **proving ground** for most of the USA's explosive weapons undergoing development. In 1984, I had the opportunity to test a Hughes *TOW-2* antitank missile, still in evaluation before production approval. This second-generation version had longer range, a larger warhead, and earlier fusing to enhance armor penetration. It is guided by means of a trailing wire keeping it electrically connected to the launcher. The gunner places optic crosshairs on the target. A bright beacon on the in-flight weapon is tracked by a sensor in the launcher, and a computer sends steering signals to the missile to assure that the beacon is always placed where the crosshairs point. My test shot was against a moving net, simulating a tank, about ½-mile distant. I placed the cross-hairs on the target and pulled the trigger. A big whoosh! Tracking the target became quite difficult, since the rocket motor plume caused some obscuration. If only there had been a crossing wind to clear that smoke. Nonetheless, I was able to hold the launcher optics without jitter for the flight duration of several seconds, in spite of seeing the target only intermittently. The missile's bright tracking beacon could be seen clearly in the launcher optics in spite of the smoke. Since the missile travels only a few feet off the ground, the most common error by novice gunners is to let the crosshairs droop, causing the weapon to strike the ground before target impact. Fortunately, this green operator still managed to achieve a direct hit. The experience not only improved my bonding with USA field commanders but also increased my acceptance by the Hughes experts assigned to this project.

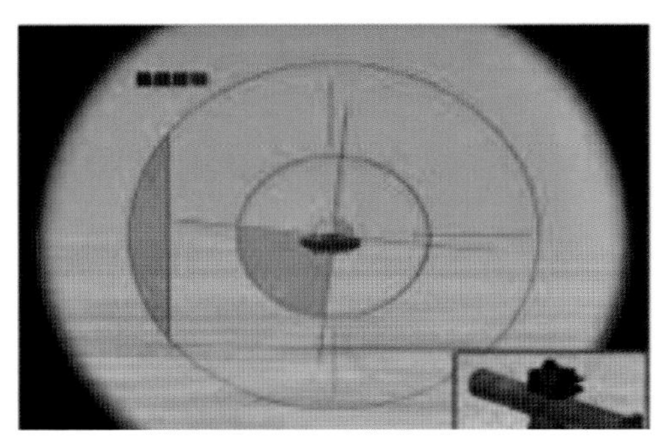

Tracking the Target

Fort Bliss is the center for **Army air defense**, with most testing done at nearby White Sands Proving Ground, New Mexico. We won a competition, with Boeing as a partner, to become the U.S. manufacturer of the unique *Roland* mobile surface-to-air weapon suite. This missile and fire-control hardware had been created and produced by Euromissile, a Franco-German consortium. A myriad of cultural difficulties in effecting this project are described in the next chapter.

The *Roland* missile system had search and track radars, an optical telescope, a central computer, electronic command links, and missiles with a ten-mile range. It was guided by radio command link from a launch trailer, which tracked the target seen by its radar or by its optical sight. The suite was packaged in modules so that all could be fitted into several varieties of armored vehicles. In the battlefield, those could keep up with maneuvering troops and defend them from marauding aircraft.

A slight misunderstanding of the *Roland* operating architecture caused a problem in validation testing at White Sands. The test director had just altered the plan for the first missile shot, specifying that the guidance would use the optical sight instead of a radar track. He also decided to leave the radar turned off. Launch was triggered, but the missile did not respond. The big surprise for all of us was to hear later from Euromissile that a hang fire would always occur if the radar was shut off. It was astonishing that the very smart U.S. team had not detected this feature previously. It was distressing for me to watch a quick sequence of four reattempted shots, all with the same ghastly result. A critical lesson was learned by us all at significant cost!

How did we get along with the U.S. government defense agencies?

Here's a positive statement by Lt. Gen. Donald M. Lionetti, former Director of the U.S. Ballistic Missile Defense Agency:

> In 1984, I had occasion to work with Hughes Aircraft to modify a *TPQ-36* mortar-finding radar for air defense use. Hughes delivered as promised in record time a prototype that was evaluated and later deployed throughout the Army as the *Sentinel* air defense radar, which is still fielded. Hughes delivered more than just technical excellence; Hughes provided top quality support to the Army with highly qualified, dedicated professionals.

Go Navy

Hughes products for the **U.S. Navy** (USN) were of every variety, covering all their many missions on the surface, undersea, in the air, and in space. Included were fourteen airborne systems; eight guided weapons of air-to-air, air-to-ground, torpedoes, and strategic intercontinental; six ship defense radars; four fleet command-and-control networks; four sonar systems; many types of large-screen display consoles for ships and submarines; and several complex satellites, some still highly classified. To fully understand the diverse cultures within each branch of the USN was exceedingly difficult. Some segments spent much effort as competitors of the others for prestige, mission dominance, or budget assignment. One fractious encounter I survived was in a meeting with a world-famous admiral of the surface forces. My foray was an attempt to garner interest in an adaptation of our *Phoenix* missile for use as a surface-to-air weapon for ship defense. After a brief overview of the idea, he shouted, "I'll be xxxxxx if we will ever buy another xxxx conversion of an xxxx air-to-air missile." That sure ended that friendly meeting.

Nonetheless, I persisted in arranging learning encounters with all these USN branches. A disproportionate number were with the air folk since my company responsibilities were dominantly in airborne radars and air-to-air guided missiles. In 1962, a great feeling of acceptance by USN Air arose when I was invited to become charter member #48 in the revered *Tailhook Society*. In addition to participating in active aircraft carrier cruises, the prominent shore-based highlights were California's Naval Air Station (NAS) Point Mugu near Los Angeles, NAS Miramar near San Diego, and Naval Submarine Base Bangor on the Kitsap Peninsula in Washington state (since merged with Naval Station Bremerton into Naval Base Kitsap).

In all the military mixing experiences, foremost was a week spent

USS *Kitty Hawk,* CVA-63
(USN photograph by MCSN Stephen W. Rowe)

aboard the aircraft carrier USS *Kitty Hawk* (CVA-63; later CV-63). At that time, it was executing strike missions against North Vietnam from "Yankee Station" in the **Gulf of Tonkin**. Rear Adm. Malcolm W. Cagle, Commander of Carrier Division One (retired as a USN Vice Admiral), and William Colby of the CIA were also on the boat. Having dinners nightly with them was most enlightening, although there were zero words about the CIA. Hughes teammate Ralph Shapiro, later becoming President of Support Systems Group, and I photographed and made detailed observations of combat operation sequences, aircraft battle damage, electronic maintenance, weapon stowage, and deck manual handling of aircraft and missile. Upon returning home, we reported a pictorial summary of our findings to three thousand employees working on the electronics and primary missile slated for the *Tomcat.* Each night it was tough for two civilians to sleep while in our bunks directly beneath a catapult as it flung aircraft aloft. During daylight, we talked to the men who were working extended hours, saw the hazardous activities on the windy flight deck, and heard of losses in some strike missions. We had been flown from NAS Cubi Point in the Philippines in a Grumman C-2 *Greyhound.* Seated facing backward, with no side windows to peer from, the 3½ hour night-time trip and the sharp jolt of the tailhook arrested landing required a lot of faith!

Our return flight was delayed due to repair problems on all C-2s in the Western Pacific (WesPac). For the next twenty-four hours, we were told to check at two-hour intervals to see whether a C-2 had arrived to fetch us back to Cubi Point. One finally arrived, but an engine failed on its final approach to the carrier. Twelve hours later (still checking at those two hour intervals), we were told to quickly board the aircraft; a repair crew had worked continuously for those twelve hours to fix one of the two engines. There had been no time to verify that repair by running the engine: scheduled strike aircraft required the deck space. An officer shouted to me, "We've got to get rid of that xxxxxxx C-2 NOW! As soon as the engines start, we'll launch!" Our return flight to the Philippines was uneventful. Unfortunately, the very next week that same C-2 had engine failure and went down at sea, losing nine USN personnel, including two with captain rank. The hand of fate had surely guided Ralph's and my return from Yankee Station.

Sea voyages on destroyers and cruisers included witnessing a successful surface-to-air missile launch against a drone. I also relished a day's cruise from Bangor, Washington, aboard a **strategic nuclear-powered submarine**. Shipping aboard the USS *Andrew Jackson* (SSBN-619) "Boomer," a Lafayette-class nuclear-powered ballistic missile submarine, was exciting: watching undersea navigation, observing simulated SLBM (Submarine Launched Ballistic Missile) launches, and vaulting upward at a 45-degree angle in an emergency breach as the huge craft literally jumped like a whale out of the sea. Our company had supplied about half of this huge boat: electronic systems, displays, advanced torpedoes, and the clever computational guidance systems for the SLBMs. It was most informative to see them functioning properly and to capture a feel for what it was to be aboard as an officer or crew member. It takes a lot of personal "sand" to be submerged for six months while awaiting a possible horrific command to launch one or more boomers.

USN Nuclear Powered Submarine

Active Aviating

Now came **combat flying**! Skilled military pilots are a special breed, with talented eyesight, quick reactions, intuitive feel of what an opponent might do next, steady and firm mental and muscular control in dynamic conditions, and ultimate calm in hazardous situations. Getting a real understanding of how they think and behave is best accomplished by "being in action" with them. Since the age of five, I had dreamed of becoming a pilot (though not necessarily in actual combat, as others did). The USN offered me an opportunity to be allowed to fly in fighters. I had already experienced many flights in seven types of military transports. Now came the real thing.

To qualify for active flight in fighters required rigorous medical and physical training. Experts at NAS Point Mugu (the West Coast center for air weapon evaluation) performed comprehensive medical exams and taught many survival lessons. The next step was exposure with seven other candidates in a vacuum chamber taking us to the extremely low air pressures existing at high altitudes. One exciting segment was to remove the oxygen mask at 25,000 feet to see how long you could perform hand motions before blackout. I carried on for about two minutes. Then, with worsening tunnel vision, I finally became unable to place playing cards properly on a table. On came that oxygen mask savior! The next thrill was checking eardrum reaction with a sudden pressure decrease of several thousand feet (simulating a sudden puncture of a pressurized aircraft); that was not pleasant. All this hazardous testing required a lot of trust in the chamber operator. Real fun was an ejection-seat trial from a fighter cockpit attached to a platform supporting a near-vertical travel rail with a stopping block at its top. After student strap-in, a small cartridge discharge boosted the seat upward, at an acceleration of three times the force of gravity (3 g). After hitting the stop, the chair slowly returned to the floor with a delighted student. 'Twas just like an exciting Disneyland fantasy ride.

Scheduled for January 2, 1963, was a lesson on how to survive a parachute descent into the sea. This was done in a large outdoor swimming pool. The pool had not been heated for the holiday week, so water temperature was about 50 degrees F. The candidate was dressed in full flying garb with heavy boots, g-suit from the waist down, a helmet, and a parachute harness. He stood poolside with his back to the water. A sturdy rope was tied to the harness where a full parachute would normally attach. On "go," the rope was rapidly jerked from the far end of the pool.

The victim hit the water upside-down and then dragged backward while he tried releasing the parachute harness. To pass the survival test, he then had to stay afloat for ten minutes. It was a soggy and uncomfortable ordeal in that frigid water. The urge to remove my heavy gear to avoid sinking was offset by the strong desire to stay warm. It was a good thing that I was already an experienced swimmer.

Now properly qualified, my first **fighter flight** was as RIO (Radar Intercept Officer) with USN Lt. Harry Conrad as pilot. We departed Point Mugu early one morning in a McDonnell Douglas F-4B *Phantom*, headed for the sea test range. As we left the runway, ground control requested us to pass close to a Chance Vought F-8 *Crusader*.

Author with USN Lt. Conrad in F-4B *Phantom*
(personal files)

Its pilot thought that his target-banner towline had not properly reeled. Lieutenant Conrad approached close from the rear and executed a complete roll around the F-8 to see if the snagged line could prohibit a safe landing. During this barrel roll, we both peered straight upward through the F-4 canopy at all the F-8's exterior. After we confirmed that all appeared OK the F-8 left for home, and we zoomed over the Pacific. Conrad performed intense high-g turns, a split-s dive, reached 55,000 feet, and zipped along at twice the speed of sound. What excitement for a civilian, especially one who luckily was impervious to nausea or altitude sickness! The very next week, Lt. Conrad and his USN RIO experienced an in-flight failure and had to eject from that same F-4. They successfully parachuted into the sea and were rescued by a helicopter. Both survived, but the RIO sustained a broken leg. 'Twas a good thing for me that, just as it had been in the Gulf of Tonkin, my flight schedule had not put me on that unfortunate later flight.

Naval Air Station (NAS) Miramar was the USN West Coast center for fighter-pilot training (that function has been relocated to NAS Lemoore, in central California), including the world-famous Top Gun School. That had been established during the Vietnam War when some bright professionals noted that all U.S. air combat training used the same aircraft types for both sides of friendly vs hostile encounters. Our success rate against the North Vietnam MIGs was not what it should have been, as had also happened in the Korean conflict. If training always pitted diverse aircraft types as opponents, the tactics and maneuvers should become more realistic and effective. That proved very true since our combat favorable exchange rates soon soared dramatically.

The Douglas TA-4 *Skyhawk* trainer allowed a student in the rear seat to operate all the flight controls. Capt. Scotty Lamoreaux, commander of this famous NAS, took me aboard one of these birds for an hour of practice over the Pacific. This backseat tourist did OK in **attempting combat maneuvers**, but performance was pretty dismal when executing a horizontal roll: we skidded all over the place! Doing it correctly requires much training in continuously coordinating movements of the throttle, ailerons, and rudder. Operating agile and powerful military aircraft requires a lot of skill and practice. Achieving success by a civilian rookie in his first attempt ain't easy. In 1959, Scotty set the altitude record for powered flight using an F-4 *Phantom*. The aircraft zoomed to over 94,000 feet. It then became uncontrollable, until it tumbled down to 60,000 feet. In 1961, he had been on a USN team that set a transcontinental speed record, also using F-4s (this was soon bettered by the SR-71 *Blackbird*). Scotty also completed over 250 fighter missions in Vietnam.

For further airborne **hands-on experience**, I had two back-seat flight adventures in an F-14 *Tomcat*. These were professionally and emotionally rewarding. Perhaps I could now feel a bit of the "right stuff." Both flights were far more than just routine observation jaunts for a lucky tourist. Important engineering tests were involved. As system operator, I had to do precise and difficult work as if in real combat: pretty challenging for this senior Hughes manager of the *Tomcat* fire control and missile. Early that morning my first "casual" flight became technically important. The previous day's flight had revealed design flaws in two captive-carry *Phoenix*. Overnight fixes had to be verified in flight before the real test shots that had to be done the following day. George Marrett piloted, and I was the inexperienced weapon control officer (WCO). We headed 100 miles inland to USAF's Edwards AFB. That area was always used whenever a Soviet electronic signal-snooper trawler was cruising the USN's Pacific Missile Test Range. If we stayed below 20,000 feet, our system radiations would be masked below the horizon, so the trawler could not capture and analyze our classified emanations.

Supersonic *Tomcat*

When we were positioned properly, I manipulated the radar to detect and track air targets at over 140-miles range. The test mission then began with tracking an F-8 *Crusader* head on at 30 miles. Upon closing to 8 miles, we turned sharply at 3 g's, and I "launched" a *Phoenix* (the weapon's electronics were active, but the missile was mechanically held to the F-14). Those high g's made my arms and head feel three times normal; control manipulation was challenging. After target crossover, we turned sharply at 6 g's (wow!) toward the F-8's rear. I then relocked the radar using a mode for tail encounters and triggered the second *Phoenix* 1 mile from the F-8. Our missile locked and successfully tracked to "impact." On both encounters, George steered the *Tomcat* to within one wingspan of the target, so the missile seeker could track almost to a combat impact, gathering data to see if the fixes worked properly. They all did, and the next day two live missile shots scored direct hits. Exhilarating for a novice to have contributed to these successes.

The second F-14 venture was piloted by Bob Solliday, a former Marine who had also been a finalist in the astronaut selections for NASA's *Mercury* Program. The mission, flying over the China Lake Naval Air Weapons Station in central California, was intended to attain data to set computer software of a new **helmet mounted display**. This device allows the user to aim crosshairs at an object, then command several systems to align to that same angle: radar, infrared tracker, laser pointer, or missile seeker. Combat engagements could proceed immediately without the crew cross-referencing on the instrument panel. Needed were accurate measures of the distortions resulting from the canopy's swelling or shrinking from altitude pressure changes. My job was to continually point the helmet-display crosshairs at an F-8 *Crusader* flying a mile abeam, as both aircraft gradually ascended from 20,000 to 45,000 feet. This task was repeated as we slowly descended. Angle data between the tracking radar and the helmet sight were to be recorded to provide real numbers to define software for computer compensation of those canopy distortions. The helmet was heavy, and holding my neck in one position for so long (about ten minutes up, then ten minutes down) was a ghastly ordeal! Any head movements would invalidate the test. Most unfortunately, no data were recorded: I had not properly switched on the recorder. The activation switch was not visible, being far behind my right shoulder. I thought all aircraft switches move forward or upward to command on; the opposite way for off. I don't remember being told before this flight that this temporary switch had been rigged backward. Rats! It was most embarrassing to have blown this expensive mission with a single goof.

All these memorable fly-boy adventures were helpful in all future customer contacts. I now could speak their language and at the same time feel extremely humble in light of what they do so well. Our firm was considered very positively for fulfilling our obligations by all branches of the USN. Here is a view by an eminent senior officer. Vice Adm. Robert Baldwin, former Deputy Chief of Naval Operations for Air:

> I remember several generations in the "good old days" of working with various industry agents. My lasting impressions of the Hughes people were that in comparison with other aerospace reps the Hughes people were more focused on the product and not remembered for drinks and dinners! I do remember your people as being technically competent and responsive to performance questions.

7

BLENDING WITH OVERSEAS CULTURES

OVERSEAS MARKETS MAY provide a bonanza of opportunities to develop and produce products. This is especially true when your organization has a world lead in some types of advanced technology. There is likely a great desire abroad to obtain unique hardware that will improve the lifestyle, economic status, and defense posture within the nation. Leaders of government and industry may be willing to invest large assets to satisfy many desires of their population. As the old saying goes, "thar be gold in them thar hills," but success in these new markets requires careful planning and preparation. It will be challenging to adapt comfortably with the diverse cultures of many overseas nations.

With the approval of the U.S., many aerospace contractors furnished to other countries complex mechanisms of many types. Hughes successfully supplied products to fifty overseas countries, with a dedication to assure ongoing maintenance support, operator training, and options for performance upgrades. Items ordered were military hardware, satellite communication systems, air traffic control networks, secure communication equipment, and scientific devices for space research.

In addition to price agreements in setting up a deal, two significant negotiation issues may be very sticky. One is coproduction, wherein the customer wishes to manufacture as much of the hardware as possible. The other is technology transfer, in which advanced design or manufacturing techniques are directly revealed to in-country organizations. Both of these are difficult: coproduction reduces the supplier's business base, and brainpower transfers may lead to future competition from a new source.

Before doing any of this, it is always paramount to respond to and satisfy any customer by acting in a personal manner comfortable to their way of operating. Some cultures are formal, others casual; languages differ both in words and innuendo; food delicacies may be symbolic; there are differences in respect for the elderly; unusual forms of humor may be mysterious; views of historic relations between nations may be emotional; entertainment styles may appear offensive; and current political leanings vary enormously. Gaining comprehensive knowledge of particular behavioral traits is necessary. This can be done by studying appropriate books, using an émigré from the particular country of interest, or by hiring a consultant. Having a friendly expert at your side in every contact can be most helpful. It is possible that you may be inclined to revise your own behavior, which may have been stilted by a life exposed only to an insular U.S. environment. A master in this trade is one who can blend well with other cultures, shows respect for history and traditions, understands needs, can adjust personal behavior to style, and has familiarity with the other language. Practice may make one perfect, if done before the important meetings. It is essential that after some rapport has been achieved or a business deal made, that prompt responses be rendered to any questions or concerns expressed by the other party. Very fortunately, in overseas transactions, many amusing incidents will occur, as each of the following eight examples illustrate.

Swiss Bliss

Hughes Aircraft Company had many important programs and enjoyed fine relationships in Switzerland. Evidence of Hughes' success in adapting to the Swiss culture and history is illustrated in this statement by the distinguished Swiss leader Lt. Gen. Walter Dürig, former Commander, Swiss Air Force:

> During the long Cold War the Hughes Aircraft Company was an important supplier to the Swiss Air Force and its other military services. Following an intensive competition with European companies in 1961, Hughes supplied the electronics and *Falcon* armament for our thirty-six *Mirage III* fighters. In the next two decades, the performance of this weapon system was one of the most impressive in Europe, and was in active use until 2003.
>
> In 1965, Hughes won the *FLORIDA* program, which included surveillance radars and a command-and-control system for nationwide air defense. It was the first computer network in Switzerland, and stayed in continuous operation until 2000. It was Europe's most effective air defense and military air traffic

> control network. The effective and high quality Hughes air-to-ground *Maverick* armed our *Hawker-Hunter* squadrons from 1982 until the aircraft was decommissioned in 1994. Following these successes many other Hughes products have played vital roles in equipping the Swiss Air Force and Army: F/A-18 *Hornet* radars, air-to-air *AMRAAM* missiles, combat training devices, *TOW-2* antitank weapons, and the night-vision fire control system for our *Leopard-2* main battle tanks. These excellent Hughes equipments have more than satisfied the military needs of Switzerland. All programs were completed, occasionally after some project difficulties, to a successful conclusion. Close collaboration between Hughes and the Swiss authorities led to many lasting friendships.

One can hardly aspire to greater praise.

Matterhorn from Zermatt

Climbing Fujiyama

Japan purchased the McDonnell Douglas F-15 *Eagle* fighter containing a superlative Hughes *APG-63* **radar**. Coproduction arrangements dealt primarily with what portion would be made in Japan. There was no concern about their ability to manufacture high-quality electronics. Industry there had a well-established record in excellent commercial electronic products. The tough discussions were on transfer of the unique technology at the heart of the radar's programmable digital signal processor, a high-speed computational device containing sophisticated and flexible software. Hughes considered it to be proprietary; all our competitors were far behind in this field and were ravenous to garner the design without making their own research investment. The USAF strongly stated its details had to be kept at home for military security reasons. Accordingly, specific drawings for that processor were not released for Nipponese manufacture, an act that remained contentious for many years. Despite that continuous dispute, the project was extremely successful for all concerned, and forty-years later this continues to be the prime fighter aircraft in Japan.

Japanese Air Self Defense Forces F-15 Fighter
(USAF photograph by Tech. Sgt. Angelique Perez)

The author's limited Japanese, learned in Honolulu while listening to a favorite radio series "Cherry Blossom Melodies," was helpful in *Eagle* negotiations. As an honored guest at a banquet sponsored by Mitsubishi Corporation's CEO, protocol called for a thank-you speech, with a translator nearby. Beginning with flowery phrases in English, I switched to Japanese. Amazingly, that translator, who did his job simultaneously as I talked, instantly switched the phrases back to English! He had been given no warning that the speaker would try both languages. That displayed immense talent and poise. The chairman responded high praise of that speech, however his associate whispered, "You did a fine job by using the female version of our language." Little did I know that, just like Spanish, this Asian language has that special ingredient! Perhaps that "Cherry Blossom Melodies" lady DJ was to blame.

Another difficult cultural experience was at a dinner meeting with that company's senior officials at their private geisha house in the Tokyo Ginza. This principal guest had been warned that his duty would be to enthusiastically proclaim that he personally savored the first dish served.

The hosts would diagnose the guest's true flavor reaction by observing his facial expressions. Fortunately, the elegant geisha assigned to me refilled my sake cup several times before that unknown dish arrived. It was a beautifully arranged artistic presentation, with flowers and green leaves in a red lacquer bowl. Atop a leaf was a complete pickled trout. Lifting it with chopsticks, I bit off the head, showing great delight while crunching that delicious morsel! Luckily, the sake had made me sufficiently mellow so that my "uncivilized Western" tummy did not react improperly. My hosts knew this show would be a real trial for me, and they politely gave a hearty round of applause for such stellar bravery.

Torii Gate at Miyajima

Understanding the customer's language can be very beneficial in establishing good relations and overcoming misunderstandings. That skill can be a valuable tool during initial negotiations, if not revealed until a deal is settled. By discreetly listening to conversations between them in their own tongue, you will get a better feel for their negotiation limits for the final contract arrangements. In the 1970s, Japan selected Hughes to deliver-in-orbit the weather-surveillance *Geostationary Meteorological Satellite* (GMS). As always, different languages and technology complexities foiled accurate communication. When our own professional language translator was temporarily absent, customer participants would exchange their personal views in Nipponese, assuming we did not understand. Harvey Palmer of Space and Communications did not reveal his adeptness in Japanese, which he had learned while on Army duty in 1946. Harvey was thus able to mold our negotiation to blend comfortably with this customer.

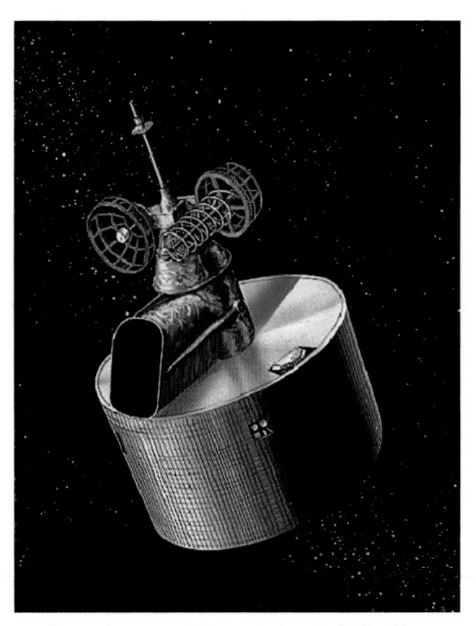

Geostationary Meteorological Satellite
(courtesy of UNLV)

Saludo a España

Another fine know-the-language situation was quite effective but had a humorous and delightful ending. In 1980, **Spain** became a sales opportunity for the F/A-18 *Hornet*.

A USN, McDonnell Douglas, Northrop, and Hughes team made final marketing presentations to persuade the Spanish Air Force to select that fighter rather than the General Dynamics F-16 Fighting *Falcon*. As Program Manager of the Hughes *APG-65* radar, it was a privilege to travel to Madrid seated next to Kent Kresa, Northrop's CEO; rewarding me with another professional learning experience. Leading this venture was USN Program Manager Admiral John Weaver, with whom I had been associated since 1968 and have always greatly admired. The task was to describe the segments of this radar that could be co-manufactured by Spanish electronics companies, and also indicate what specific technology could be transferred. On the last evening, the Spaniards gave an elegant dinner for us all at Madrid's outstanding Zaragosa Restaurant, a place

Cibeles Fountain, Madrid

with Renaissance-era decor and costumed waiters. Hosts and Americans were interspersed. Being next to the Deputy Minister of Defense, I complimented him as a look-alike with King Juan Carlos (a dead-ringer). Then I spoke several sentences in Spanish jargon that had been acquired over the years. In the four days of discussions, no indication had been given about this rudimentary understanding of Espanola words. It had been possible to understand many phrases that these potential customers had assumed were private. The Minister beamed, raised his hand for attention, and announced, "I caution my countrymen: Mr. Richardson speaks 'Mexican' Spanish!"

First Production F/A-18 *Hornet*
(USN, Naval Aviation News)

He scored a perfect put-down; the upper crust in Spain use only the refined Castilian. Catalan, used by one-third of the country's population, is scorned by the privileged ones and Mexican dialects are considered grossly primitive. Those elites apparently don't resonate to the color and dash of the savory Mexican flavor.

Viva Mexico

However, in 1990 that language style came in handy in a meeting with President Carlos Salinas de Gortari in **Mexico City.** The visit's mission was to persuade him to overturn the actions of his federal procurement director.

Teotihuacan Pyramid

At issue was the selection of a **synchronous communication satellite** to service the entire nation. A French outfit was our competitor, subsidized by its government. In contrast, our offer had no U.S. funding assistance, nor any endorsement of the product. The resulting lower French price made their bid attractive, but the satellite offered had very limited capability compared to our *HS-367.* Three times a government technical review team had selected Hughes as the winner but had been overruled by the procurement chief. Rumors abounded that perhaps someone was offering him rewards if the French system was selected. El Presidente was most impressive. He had been Harvard educated, was able to comprehend technical jargon, and spoke flawless English. As one already very fond of Mexico, I felt relaxed and comfortable to be part of this important discussion. Attempting to attain rapport, I tried my jumbled Spanish and was warmly praised for trying. Falling back to American English, several major performance and operational advantages that favored our offer were cited. Then I revealed a persuasive possible use of *HS-367*, not previously mentioned in this competition. This fresh concept popped into mind on the flight south while planning what to say that afternoon. Instructions by a single excellent teacher can be relayed by our satellite and captured by all elementary schools throughout the nation, each having a simple receiver and display. As a new enhancement of our offer, three of these classroom receivers were added, at no increase in total price. This easy-to-implement idea to improve national schooling did the trick. President Salinas ruled the next morning that Hughes was the choice in this competition. That educational boon was successfully carried out in the next few years. It continues to feel good that an idea that magically appeared one morning not only helped capture an important piece of business but also became a major improvement in Mexico's educational network. Regrettably, this El Presidente was subsequently accused of malfeasance during his presidency and exiled to Ireland.

HS-367 Synchronous Communication Satellite
(courtesy of UNLV)

Tour de France

Because of our product line array, direct sales to **France** were rare. That country is proud in self-regard, has many organizations with widespread technological capability, and can be quite nationalistic in economic and business matters. United States companies at the forefront of electronics must always be wary of inadvertent technology transfers. That concern makes business dealings always tinged with caution. Newcomers in any significant dealings must be alert, not always using the casual mannerisms usually found in America.

Notre Dame Gargoyle, Paris

A big learning experience happened in our *Roland weapon system* slated for the USA. The equipment was designed and manufactured by Euromissile, a team of French Aerospaciale and German MBB companies. The hardware for our Army was to be produced in the United States: Hughes manufacturing the electronic weapon control systems and the missiles, Boeing producing the armored vehicle. Technology transfer was to travel from Europe, compensated by negotiated fees, though many alterations were needed to match American performance changes and domestic manufacturing methods.

It was soon discovered that the French are magnificent hosts, especially in business situations. Beginning in 1975, enjoyable gatherings were used as a negotiation strategy not obvious to Americans. Luncheons at an elegant restaurant in the Aérospatiale's headquarters near Paris were extravagant: three-hour repasts enhanced with five varieties of wine.

The host's practice was to raise important issues for discussion and then ask for agreement just before dessert. Naive U.S. participants were frequently disarmed and casually agreed to contract changes, instead of taking vital issues under advisement until the next morning. These encounters were very amicable, but the French always reaped any resulting business benefits. German team members were also present; they had already experienced this ritual for three years and had learned restraint. When we began to attend, the Germans were quite amused, but they soon taught us newcomers how to cope with this charming but tricky segment of European culture.

Roland Missile System

Communications involving a complex product like *Roland*, designed overseas, was tough. Cultural and language differences caused a six-month delay in completely transferring the necessary knowledge. Euromissile sent a senior engineer whose native tongue was French. When asked to prepare an information package within two weeks, he said, "d'accord de principe." Dr. Elliot Axelband, our Program Manager, assumed that meant, "Yes" (a literal translation is "of agreement in principle"), but the phrase actually is a polite Gallic way of saying, "No." This lead man did nothing before he next appeared. Sometimes vital technical information was given only after we agreed to a license fee increase. Of course, major effort was needed to convert all measurement from metric to English units, to match our machine tools and employee operating style. If only the U.S. would switch everything to metric, we would benefit in many ways! Another oddity was the manufacturing procedure phrase "just see Otto." That Teuton was the only European capable of fine-tuning a critical component prior to final assembly. What he did could not be expressed in words. Since we would not have Otto with us, we had to create new instrumentation as a substitute for his wisdom. The U.S. *Roland* turned out extremely well, despite many traumas, three-country cultures, and our initial weakness at the Paris lunch table.

Persian Delights

Iran was a most unusual encounter. Although the language Farsi is closely akin to English, our company had few experts in that tongue, or in the ancient country's very different culture. Attitudinal differences are dramatic. What formerly was Persia is fully steeped in folklore and Muslim heritage. The nation is rich with historic and classic treasures, especially Isfahan, Persepolis, and Shiraz. During the 1970s, Iran purchased seventy Grumman F-14 *Tomcats*, containing the Hughes *AWG-9* weapon control system and the multi-shot *Phoenix* missile. This business arrangement did not involve any in-country manufacturing or technology transfers. The major daily challenge to all the Americans was to somehow find

The Gate of All Nations, Persepolis

amicable ways to blend with the significant behavioral characteristics of these people. This was a sharp contrast to the relatively comfortable adaptation to European cultures. The squadrons were to be based at two new airfields. To help in the complex field implementation, Hughes sent two hundred support experts, under the leadership of Ralph Shapiro. They were tasked to train Iranian Air Force (IAF) personnel to operate and repair the weapon control system and its companion *Phoenix* missile.

Many of the trainees had a working philosophy very divergent from ours, and those differences were almost impossible to overcome. Attempts to repair an electronics malfunction were often thwarted by the local belief that the failure was "an act of Allah's will," so must not be tampered with. A difficult mission on one of my visits was to provide analytical data to the Air Force command General Mohammad Amir Khatami on how many *Phoenix* missiles should be procured. Our staff had prepared a graphical calculation method with interactive lines of variables plotted. Tracing lines step by step to sequential intersections, a final quantity could be obtained. Desired positions along those lines were chosen by the IAF. Important variables were the number of potential hostiles, likely time to rearm combat squadrons, probability of successful intercept, and total cost. Strategic reasons for the *Tomcat* purchase were to defend

Iranian *Tomcats*

against an Iraqi threat and to deter further Soviet surveillance over-flights.

General Khatami applied the graphical method, yielding six hundred missiles to purchase. This general was a most impressive leader; talented and well-rounded in most subjects; but oddly, we had forecast that six hundred before this visit to Iran. The positive outcome of this graphic tool was that IAF made the decision on the quantity it wanted; politically sound compared to our promoting a number that would please us. Tragically, he perished shortly after our meeting in a recreational hang-glider crash. Rumors circulated that a wing-brace wire had been partially severed prior to his recreational flight by a dissident subordinate. Khatami was replaced by General Amir Hossein Rabii, who had been commander of the IAF's squadrons at Shiraz Air Base. This fine man taught me many things about the fascinating Persian culture. A special treat was to enjoy the national dish, chelo-kabab: a rice mound overlaid with raw egg and skewered lamb, accompanied by a pita called pebble bread (sangak bread). He was most likeable, and it was very grievous when worldwide television showed him being executed by a firing squad during the Islamic revolution. According to an IAF report, sixty *Phoenix* missiles were used in the eight-year war with Iraq, with great success. They even claimed that two opponent aircraft were downed by only one *Phoenix*.

Dome of the Rock

Business with **Israel** was also a revelation of differing cultural and operating styles. That nation has a wealth of technical talent, but is always primed to acquire new discoveries made by others using friendly conversation rather than by negotiating technology transfer fees. In-country production is also at the forefront in any hardware discussions. Many U.S. items of military equipment, often with their design details included, have been purchased or given to Israeli combat forces. American defense contractors, including Hughes, performed for them a great deal of consulting and initial training. There also were opportunities to market improved designs or completely new products, if given the OK by the U.S. Government.

There was an opportunity to visit Tel Aviv to try to excite the Israelis to purchase *Tomcat* fighter systems, just as Iran had done. It was astonishing to see their intense zeal for information. There were many well-informed technical questions and extensive probes. Most were focused on how to defeat the *Tomcat* by tactical maneuvers or with electronic countermeasures. It was obvious that their main interest was not in a purchase, but rather in ways to oppose Iran's *Tomcat* fighter squadrons.

To their dismay, U.S. security rules helped to keep me pretty mum. The meeting revealed very bright technical minds in these Israeli Air Force participants. It was amazing to see junior officers frequently interrupting, contradicting, and debating with their senior officers, including their commanding general! Allowing this casual disregard for rank can be lauded since information exchange can be comprehensive and junior officers will be highly motivated by this chance to show their knowledge. In most military outfits, even foolish orders must be carried out without question. It must be tough to know when to speak or when to smartly salute.

Dome of the Rock, Jerusalem

Always be prepared to be flexible in negotiations; some fair compromise may be reached with persistent, nonaggressive efforts. But be ready to stand firm when your limit has been reached. In 1988, Hughes was marketing the *RealScene* system to Israel. This equipment provided full-color three-dimensional dynamic images of target areas to help strike aircraft pilots prepare a correct approach for accurately delivering guided weapons. In this potential purchase, all contract terms between our company and Israel were agreed upon, except the final price. A senior director of that country's military procurement approached me to arrange closure. We had priced the effort with some margin to cover design uncertainty since the system was not fully developed. A very modest profit was also included. This official wanted—virtually demanded—a substantially lower price, saying, "This is for the good of Israel, so you should be willing to accept a lower price." He also hinted that U.S. Government pressure could be brought to bear on Hughes unless I conceded.

I offered this quick response:

> Although it may be for a great cause, the money you will be using for this purchase will come from the annual U.S.-to-Israel aid package. Those funds are partly drawn from federal tax payments made by this company. This program will cost your country nothing. It's already a gift from the United States, and our company does not wish to add a further donation by assuming the entire contract risk.

He was infuriated, but his unsuccessful ploy seemed a bit outrageous to me. The contract was awarded the next day at our price. After it was delivered, *RealScene* was applied very effectively in Gaza against Fatah. Militant bosses were housed in an upper-floor suite in a high-rise building. The pilot correctly delivered a guided missile directly through that room's window with all those leaders present, and there were no civilian casualties. The delivery approach, skirting other buildings, had been well practiced on the *RealScene* trainer. Our price was likely then viewed as a real bargain.

It was delightful in the 1990s to meet Maj. Gen. Yeshayahu Gavish, who had led the Israeli tank force through the Sinai Peninsula in the 1967 Six-Day War against UAR (United Arab Republic: Egypt & Syria), and Jordan. That advance, against a much larger armored combat force, was quickly victorious and the Israelis soon crossed the Suez Canal.

Precision Strike in Gaza
(photo by Yaseer Qudih/Associated Press)

That step caused international pressure for a cease-fire. Most remarkable was the rapid destruction of the many Egyptian Army's Soviet-built tanks. This apparently resulted from the Israelis' tactical genius and superior equipment, including the Hughes tank fire control systems and *TOW* anti-tank missile.

A congratulatory remark for setting a world-record time for eliminating an opposing tank force was, "It was amazing to hear that 370 tanks were destroyed in only one day!"

With a large, proud smile, he modestly replied, "Oh no, it took two days."

That's genuine chutzpah!

Bali Hai

Sometimes marketing may involve a country's political head. Adequate cultural preparation may be sufficient, but some encounters may demand a quick and correct response to a big surprise. On one such visit to **Indonesia** to help sell the *Palapa* communication satellite network, our team, led by Tom Carvey, upon arriving on schedule, was told that the meeting was delayed two days. The Premier had ruled that no negotiations could start until the Hughes people went to a distant town and returned via a test flight of a newly designed small airliner. This was the first airplane ever constructed in Indonesia. Our team's apprehension about surviving the flight lessened a little when they stepped aboard and met a senior U.S. FAA inspector, who had been asked to evaluate the craft on this same flight. If this experienced expert was willing to risk his

Rice Fields of Bali

life, why should we be queasy? Requiring such a flight seemed a strange precondition for marketing a completely unrelated product. Perhaps the Premier wanted to advertise the successful introduction of airline service for traveling foreigners. Fortunately, the Palapa sale went through, with lucrative benefits for everyone.

Roman Adventure

And then there was **Italy,** the marvelous land of enthusiasm and joy of life, with treasure troves of meaningful historic sites. With the USA promoting it, the Italians decided to purchase the *TOW 2* (Tube-Launched Optically-tracked Wire-guided, model 2) anti-tank missile. Just as in our Japanese and Spanish deals, portions of the weapon had to be built by in-country companies, including technology transfers. The final settlement choice remaining in 1983 was to pick the Italian firm to manufacture the rocket motor. Two very competent Italian contractors vied for the job.

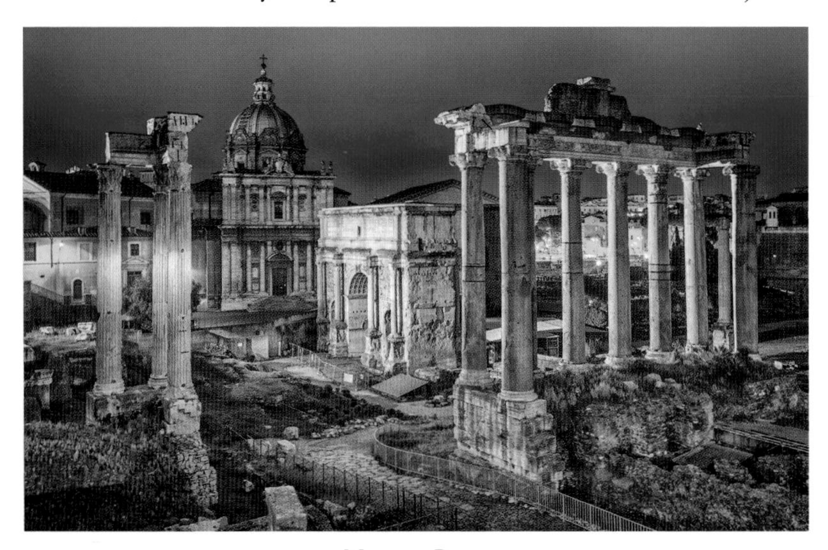

Magnetic Rome

Shortly after being appointed President of the Missile Systems Group, a strong request came from our marketing staff to give a presentation about our air-to-ground *Maverick* missile at Rome's Villa Borghese. It was expected that the audience would be a diverse gathering from many nations, including the U.S.S.R. This would be a good opportunity to entice new buyers from friendly nations; such sales being highly endorsed by the USAF. This rookie corporate executive was a bit ignorant of this product and its military security classification limits; that required lots of boning-up. Those Soviets would surely like to gather the performance limits and design details of this device.

By surprise, after the presentation there was an urgent summons to meet with the Italian Army general in charge of all national military procurement. Discussions were held the next morning in an elegant Renaissance building with high ceilings, low lighting, and loud echoes. After a half-hour of social chatting, in swept a most handsome, debonair, white-haired executive with a flowing cape and sash. He was the CEO of one of the two subcontract contenders, and forcefully asserted, "You must select our company or the *TOW* purchase will not be made by our government." This bold demand was then endorsed by the "unbiased" procurement general! Dr. Bob Roderick, our *TOW* product expert, whispered to me the current status.

With continuing astonishment and a deep inner concern that I might jeopardize the whole multimillion-dollar procurement, I responded, "We have agreed to supply to your nation the total weapon system at a fixed price, but we do not yet have a subcontract quotation from your company. We cannot select a supplier until we can compare your price with that of your competitor. We cannot bear that risk." The meeting was quickly adjourned, and the elegant CEO angrily and dramatically stormed out. It was like a scene in the movies.

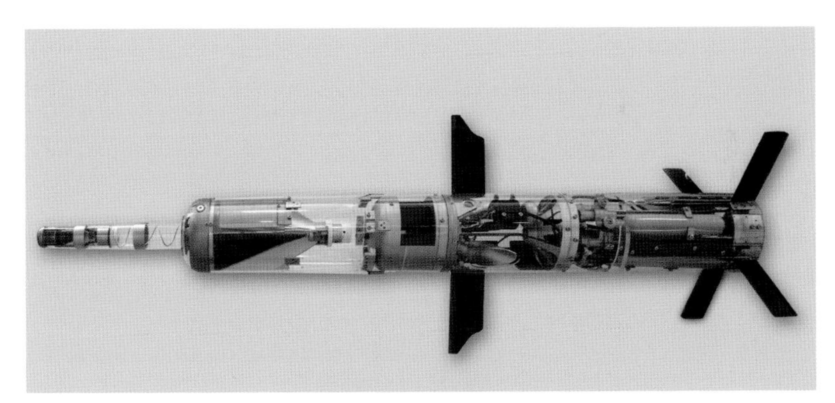

TOW-2 Antitank Weapon
(courtesy of UNLV)

Late that night a telephone caller told me to be ready at seven o'clock in the morning for further discussions. Promptly at that time, a long black limousine, escorted by several motorcycle police, appeared in front of the hotel. Upon entering the vehicle, I was seated next to the Italian Minister of Defense, Lelio Lagorio. He apologized, saying that, because of a schedule to appear that morning before Parliament, this was the only chance that he could meet with me. He asked for details of the previous day's encounter. There was no language barrier here; this government minister was fluent in English. The entourage raced down the crowded Via Veneto, with the motorcycle police clearing the path, sirens screaming. What a thrill to be doing business in this dashing manner! As we reached the Parliament building, the minister exited with a graceful bow and a warm *grazie*. The next day, the full purchase contract was signed, with the exclusion of that CEO's firm. We were comfortable with the other motor supplier's price and quality. Thank heavens we had not been buffaloed by such outlandish bombast. Who knows what happened to that officer who had been in charge of all military procurement? Let's hope justice was done.

These examples are only a few of many colorful adventures. Successful dealings with fifty nations requires much foresight, practice, grace, agility, and excellent products. Details of the large variety of sophisticated devices and services, as well as the management styles employed by the world's leading military electronics corporation, can be found in *Hughes After Howard*, published by Sea Hill Press.

8

COMPETITION FOR SURVIVAL

MONOPOLIES IN BUSINESS endeavors usually have the image of overcharging and being unmotivated to develop and distribute up-to-date hardware, software, or services. In many historic cases, that has happened. Their leaders may have preached, "If our products sell well, we don't need to spend money to improve them or to reduce their cost or price. Just keep producing them, and we will continue to profit." Such companies will feel unchallenged by competitors with modern devices and lower prices. The **monopoly** has the market power to squeeze interlopers away from consumers. However, here are two examples that had much more wholesome philosophies.

Already mentioned in chapter 4 was AT&T, established in 1881. In the first half of the next century, that conglomerate had a complete monopoly on all distant communication in the United States and some overseas areas. An enormous influx of revenue came from millions of users who purchased hardware, paid monthly fees, paid repair charges, and acquired mandatory and optional upgrades. The management decided to reinvest much of this stream of income to support and energize Bell Laboratories, one of the finest research centers in U.S. history. Its staff professionals were unconstrained by existing AT&T product lines and could pursue fruitful concepts anywhere within the diverse fields of science. Thus, many discoveries not only improved communications but also created other benefits for all of society. It still seems regrettable that the government wisdom in achieving this monopoly's

breakup in 1982 did not also find a way to counter the negative secondary effect: there now was insufficient funding flow to preserve that free-wheeling Bell Labs.

Microsoft Corporation (a compression of "microcomputer" and "software") was started in Albuquerque, New Mexico, in 1975 by Bill Gates and Paul Allen. As its need for software designers exploded, the company moved to Bellevue, Washington, where more candidates were available. The ingenuity of those two sparkplugs and their staff, supplemented by a business partnership with IBM, led to adapted forms of Microsoft BASIS becoming the primary programming language for IBM and the emerging super market for personal computers. Even its eventual fierce rival, Apple, used that software for their initial PC devices. Familiar Microsoft products today are Windows, Word, Excel, Office, Xbox, and Bing. For many years, this organization had a virtual lock on the software market, capturing 90 percent of the available business. Much like AT&T, the wise executives poured a large portion of incoming revenues into R&D, although focused on their specific product lines. Their devices and software are modernized and upgraded almost daily. Once again monopoly helped establish this giant, and only the substantially different design concepts by Apple have changed the game.

This enormous income from a myriad of consumers enabled large R&D investments to keep the technology advancement active. Apple increased its marketing position by excelling in display techniques that made individual users more comfortable with understandable operating controls. Competition is thriving today. It is difficult to find fault with the Microsoft monopoly since all of society has savored the results.

Competitive Spirit

The essence of business survival and corporate growth in our economic system is successful and **perpetual competition**. Constant effort to capture sales spurs organizations to stimulate creativity, take risks, invest in the long term, and perform with alacrity. The leadership credo might be: "If we don't continually reduce cost through productivity and factory improvements, the competitors will beat our market position. We also must always search for better product performance and new product lines, or we will be surpassed by other entrepreneurs. We must not become obsolete like automobile steam engine manufacturers."

Such drives will motivate high manufacturing efficiency and reward effective cost reduction. But they also spur R&D investment to stay ahead in device and software designs, as well as to discover new product thrusts. The challenge will also inspire the company staff to do their work with ingenuity, excellence, and efficiency.

A company competing for selection by a single customer, the U.S. Pentagon, is faced with very different challenges. These include one's reputation for fielding worthwhile equipment, ability to find breakthrough technology to meet new performance goals, efficiency in reducing the cost of development and production, and ability to nurture resonance with the using customer. Frequently, politics become crucial: Congressional budget approval is affected by where the work will be done since Congressional members are often trying to improve their state's employment. The following observations stem from personal experiences in the aerospace industry.

Best Performance Usually Wins

A development project Hughes won with a requirements exception was the *APG-65* radar destined for the USN F/A-18 *Hornet*. McDonald Douglas, the aircraft's prime contractor, evaluated bids from us and the very competent Westinghouse Corporation. Many new features we proposed, including the world's first programmable signal processor. This device enabled detailed interpretation of radar returns at extremely high speed using digital computer formatting.

Critical to any radar's performance is the measurement of antenna pointing angles within its scan limits during all aircraft flight conditions. Accuracy budgets are assigned to the many devices in the antenna chain, but regardless of the radar's precision, the angle is strongly affected by the aircraft's surrounding fiberglass radome nose cone. We noticed that McDonnell Douglas had allotted the entire beam transmission error budget to the radome designers. This meant that our engineers would have no precision inaccuracies. Since this would be impossible to do, we could not, with technical or business integrity, submit our bid. Our project's technical leader and I made a trip to St. Louis. In a formal meeting with their *Hornet* leaders, we stated that Hughes would not bid the entire program unless a reasonable error margin was assigned to our engineers. This was very risky since some competitors pledge specification conformance hoping for a contract modification after the award. The good news came several days later: we were given some relief. Even though the margin granted was very tight, our designers met the challenge. This radar became the largest production program in Hughes history. *APG-65* and dramatic upgrades are still being produced and deployed by many countries forty years after our development award. A hazardous bidding gamble paid off well.

Hornet Radar Maintenance
(courtesy of Raytheon)

Our largest production of thermal imaging devices was **tank warfare** systems, following a difficult competition in 1973. Texas Instruments had been more responsive to the USA laboratories' goal of modularity units (subsystems that can be interchanged to form alternate systems for different missions.) They had captured the initial development contract. However, we prevailed in later bids and won most of the USA production procurements. Using the unique integration of imaging infrared and laser pencil beams, these systems solved the ever-existent problem of tanks using a cannon against a maneuvering small target while racing at 30 miles per hour over rough terrain. In the World War II North African desert campaigns, a critical element in tank combat and survival was the speed achieved in finding the opponent, measuring its range, determining the cannon elevation, forecasting angle changes from maneuvers, and firing at the correct moment. Seconds made the difference, but even skilled gunners needed a half minute to measure range and shoot. Success at night or in fog and smoke was virtually impossible.

The new *TFCS (Tank Fire Control System)* was gimbal mounted and teamed a high-resolution thermal imager with a telescopic optical tracker, a laser coupled with a computer, and an operator's display. The tank gunner only needed to view the terrain ahead through the telescope, choose a target, select a weapon, and pull a trigger.

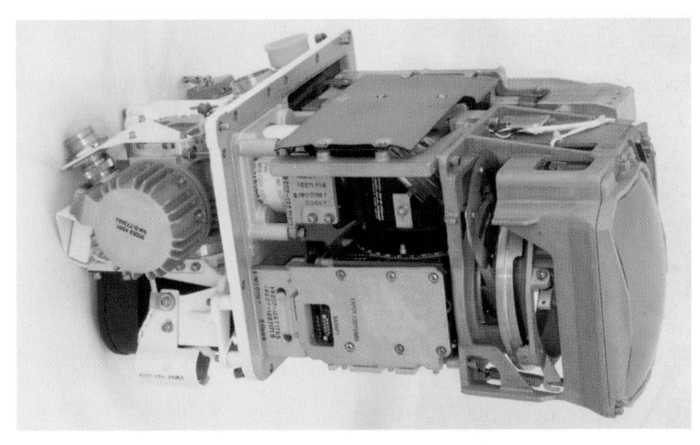

Tank Fire Control System(courtesy of UNLV)
(courtesy of UNLV)

The system's fire control then automatically compensated for all motions, kept the laser range finder pointed at the target, measured and predicted both the target's range and angle, computed the cannon shell trajectory and expected impact time, and fired the weapon. Most astonishing in our design was the stabilization system that kept all sensor devices aligned as if the tank was motionless, regardless of speed or terrain roughness. Functions were done with such precision that first-shot destruction of the opponent was almost a certainty. Those deadly time delays in World War II battles were no more. The *TFCS* gave tank crews day and night visibility through haze, smoke, and dust never before possible in the battlefield. Armor protection was maintained for the friendly crew since viewing was through a periscope. Another step forward was use of a laser beam to spot illuminate the target for the newly developed Rockwell *Hellfire* guided missile. We produced more than 15,000 *TFCS*.

During the 100-day Desert Storm, *TFCS* were employed in the A-1 *Abrams* tank and the M-2 *Bradley Fighting Vehicle*. Many times, allegedly, Iraqi tankers would immediately abandon their vehicles if they saw their commander's tank destroyed. They would leap from their tanks and surrender to helicopters flying overhead in the hope of being safe, even though it meant being captured. Commanding Gen. H. Norman Schwarzkopf Jr. remarked:

> Our tank sights have worked fantastically well in their ability to acquire targets through all the dust and haze.

There were numerous competitors for surface-based radars used for air defense, ground combat, air traffic control, and ship navigation. The big players included General Electric, Motorola, Philco-Ford, Raytheon, and Westinghouse. To gain an edge, Hughes spent three years of IR&D funding to conceive and perfect an antenna that could sweep beams without physical motion of the hardware. For the first time, a narrow pencil beam could be instantly moved from one view to another widely spaced in directional angle, using **electronic scan**. The antenna face is a flat aluminum panel with many carefully spaced slot openings that can pass microwave energy. Beams are shaped and steered by altering the operating frequency or the phase of signal streams passing through each slot (see chapter 2).

We won many new programs from all branches of the military service. Today, most USN ships employ sets of these radars for managing Task Force ship and defensive aircraft placement, as well as for detection and tracking of incoming strike bombers and sea-skimming cruise missiles.

The USA and USMC actively employ two varieties of our *FireFinder* electronic-scan radars to protect ground troops. These machines instantly detect and track any incoming **cannon** or **mortar projectile**, compute its origin, and automatically point and shoot accurate return fire. These were

FireFinder, Electronic Scan Radar
(courtesy of UNLV)

so effective in Desert Storm that Iraqi artillery crews refused to fire their cannon when commanded by their officers since that would place them immediately at hazardous risk. Similarly to *TFCS*, many thought that *FireFinder* also became a meaningful combat deterrent. Better to have the hostiles to flee than continue combat.

Buying-In Pays Off

For decades there were cataclysmic contests with Raytheon in numerous fields of combat equipment, many in high-performance guided missiles. (To entertain our staff, I referred to that outfit as "the *R* word.") We usually won most development project battles based on more advanced technology. They advertised lower production prices. The firm had an internal motto: "Whatever Hughes designs, we will manufacture." In spite of lower total cost claims to the DoD, we captured the most formidable and prestigious contest between us: the USAF program to create the *AMRAAM* (Advanced Medium Range Air to Air Missile). We also won the bulk of production quantities for six years. Again, technical leadership prevailed, especially when we underbid the development program.

The requirements were very tough: half the weight of *Sparrow*, include many alternative guidance methods, achieve longer intercept range, and enable four birds simultaneously in flight against four separate hostiles. Not only were many inventions needed with a constrained development price, but the winner would have to divulge all details to the loser.

F-22 *Raptor* Launches *AMRAAM*
(USAF photograph by Master Sgt. Michael Ammons)

That would allow two viable manufacturers to compete for portions of annual procurements. Hughes CEO Dr. Allen Puckett viewed this as a "lifetime business opportunity," so he set a bidding fixed price of $421 million. It was then estimated that completing resulting contract award might require as much as $45 million of company funds. Our business history showed that, on average, we lost 2 percent on a development program, not 10 percent. This was offset by gains of about 6 percent from resulting large production contracts. To garner the win, we promised new creations: computation speed and capacity, superb in-flight data linkage to the fighter, exotic construction materials, ablative skin for thermal protection, GPS tie-in, and an advanced design rocket motor—all with a high-power transmitter featuring our unique TWT. *AMRAAM* was indeed a lifetime program. For thirty years, it has been the primary air intercept weapon deployed by all our allied countries. The 1991 Desert Storm conflict gave positive proof of our design: twenty-eight hits scored for twenty-eight shots. Just like *TFCS* and *FireFinder*, this Hughes weapon surely became a combat deterrent to the Iraq pilots. If it were known that any fighter capable of launching *AMRAAM* was within one hundred miles of their base, they would refuse an order to take off.

In the late 1970s, the U.S. urgently needed an effective weapon to counter the very sophisticated Soviet Alfa-class submarine, with substantial performance improvements compared to the USN *Mark-48 (MK-48)* torpedo already deployed: higher speed, operation at greater depths, homing at twice the range, immunity to countermeasures, and far greater lethality. Our demonstrated expertise in intricate system engineering, creating many types of smart missiles, and developing sonar surveillance systems, coupled with a very optimistic price (with no margin for uncertainties, which always do occur!) won the competition for this exotic *MK-48 ADCAP* (Advanced Capability) torpedo. Our opponents were formidable: Westinghouse and Alliant Systems. The initial project was for the acoustic transmitter and receiver, signal processing, and a computer generating real time control of the torpedo after it was launched. The current *MK-48* prime contractor, Gould Industries, won its bid for the propulsion upgrade. Soon after our award, the USN directed that we must become the prime contractor for *ADCAP*, replacing itself in that position. Our corporate management was extremely uncomfortable to undertake this risky venture: a completely new product thrust not previously thought through. Nonetheless, in response to customer desire, we entered an offer as prime, after wisely deciding to team with Gould, which had been the prime contractor for the older *MK-48*. Also promised was a plan to build a new manufacturing plant in Forrest,

Maintenance of *ADCAP*
(USN photograph)

Mississippi, (politically beneficial in the competitive win). The technical challenging goal was to ensure high hit rates against deeply submerged nuclear-powered submarines and high-performance surface ships using evasive tactics.

Ensuing technical hurdles and significant USN changes in requirements caused continuing renegotiation of price adjustments to remedy substantial financial shortfalls. The USN budget relief requests caused the Congress to send several inquiry teams to Hughes. Fortunately, they agreed strongly with our position and were impressed with our candor and honesty. The resulting successful weapon was 19 feet long, 21 inches in diameter, and weighed 3,695 pounds. Midcourse guidance was by commands sent from the launching submarine through a de-spooling wire (similar to our *TOW* concept described in chapter 6) or by semi-active sonar echoes bouncing from the target. Terminal guidance used an active sonar mounted within the torpedo. First deployment in 1988, more than 1,000 had been made by 2007. This success led in the 1990s to further expand torpedo production to include the *MK-46* and *MK-50* lightweight torpedoes by acquiring the Marine Systems Group of Alliant Corporation. What began as a great uncertainty became a long-term winner.

Ferocious Confrontations

The AIM-7 *Sparrow* missile, designed and produced by Raytheon, had been the only radar-guided air-to-air missile for the USN, and had been somewhat successful in the Vietnam air battles. For the F-14 *Tomcat,* it would still be one of its weapons, but the Hughes AIM-54 *Phoenix* would now be dominant due to its long range and the unique ability to have six in guidance at the same time against widely spaced targets. At stake for the two corporations was the total quantity of their type to be procured. Raytheon would be delighted if, somehow, the primary armament for *Tomcat* became *Sparrow* rather than our bird.

USN *Sparrows*
(USN by Photographer's Mate Airman Tommy Gilligan)

An important flight test objective was to assure safe separation of a missile during launch while the aircraft is executing sharp turns. A really tough one is when a shot is tried while the fighter is at a most-unlikely combat maneuver of **negative 3 g's** (-3 g) (the pilot commands an extreme downward push-over rather than the traditional pull to execute tight turns). The launcher device must deliver enough push to counter a strong reverse gravity force that is trying to shove the weapon back against the airframe. Raytheon agreed with the USN to allow this test for its bird, but I demurred for a *Phoenix* launch. Our analysts were convinced a safe separation would not occur. The project's lead admiral in the Pentagon was most dismayed and had some strong words to me demanding that I should change my decision. It is almost mandatory never to antagonize the customer, but this was a significant safety issue. Our competitor saw this as a strong aid to persuade the USN to re-designate what was to be the fighter's prime weapon. Everyone was most aghast when during that -3 g test, the *Sparrow* remained close to the *Tomcat* fuselage, the rocket motor ignited, and its vertical-facing wing fin sliced a long line through the fighter's skin. Spilling fuel ignited, and the F-14 was lost at sea. Fortunately, both crewmen escaped. Which company did the right thing?

In another confrontation our adversary asserted to the USN that the *Phoenix*, which in midcourse only travels at Mach 3, was woefully **less maneuverable** in the terminal phase than *Sparrow*. It speeds in midcourse at Mach 4. We examined all the physics involved and concluded that since our bird was tail-control steered, its flight speed was little lessened by increased drag from steering. In contrast, the *Sparrow* is steered by turning its centrally located wings. That generated far more drag, quickly reducing flight speed and maneuvering energy. To counter our competitor's claim, I visited Washington to present our analysis to Naval Aviation's prime technical chief, Dr. John Rexroth. Our one-on-one chat was most successful, and this suggestion by our opponent to cancel the *Phoenix* was rejected. Surely is nice to be allowed to reveal the true facts in any dispute.

My most bitter political influence event occurred when Raytheon was being paid by the DoD to establish a production second source for *Phoenix* missiles (see "Government Handicaps"). We were required to educate them on all details and to provide experts on their site to answer questions. One assembly element was security **classified "secret."** Rigorous and costly special compartments with watch guards were mandated for its assembly on the factory floor. Our on-site Hughes staffer noticed that no such provisions were being installed at the Raytheon assembly line. I called this to the attention of the appropriate USN officials since it was a sharp security violation and would also allow our opponent to operate at lower cost than we. For some reason unknown to me, the next day Raytheon was excluded from that construction requirement, but the burden was not lifted from us. Could that possibly have happened because our adversary was in Massachusetts with a strong Congressional supporter to redirect the USN? Who will ever know?

Political Geography Helps

A dicey concern in competitions for Pentagon-solicited projects is the possible **influence of politics** in the final selection process. Defense budgets are set by Congress, and many members are strongly biased toward companies within their district, or wish to curry favor with other influential people. They or the Administration can place strong pressure on the DoD staff, with the implication: "If you do not pick the corporation I designate, regardless of proposal evaluations, I will not approve your budget."

USAF Flying Wing Bomber
(USAF photograph)

A classic example occurred in 1947 when Secretary of the USAF Stuart Symington cancelled Northrop's JB-1 *Bat* flying wing program. (It was to employ the world's first radar guided missile, the Hughes JB-3 *Tiamat*). A flying-wing aircraft has no rear boom to support rudder and tail control surfaces. Mr. Symington wanted JB-1 production to be done in Texas rather than California. Since Northrop refused to do so, the program and all successful prototypes were destroyed. It's ironic that *Bat*'s ultimate successor, the B-2 *Spirit*, an excellent stealth bomber, was produced in California by Northrop and successfully deployed in the 1990s.

Hughes both lost and gained in competitions because of this workforce and facilities location decision criterion. Possible losses described in this chapter were the FAA *Air Traffic Control System* and *THAAD.*

We built and staffed plants in thirteen states outside of California. Many programs greatly benefited from these placements, enjoying lower costs and higher quality, as well as those political advantages compared to competitors. Most notable were *ADCAP* torpedo in Forrest, Mississippi; *Tank Fire Control System* in Orangeburg, South Carolina; and *Maverick* missile in La Grange, Georgia. These sites included fully equipped plants for sophisticated electronics manufacturing. Those employees displayed high motivation and a desire to excel: this level of job was far better than any other in their or nearby communities. The high motivation of employees to doing our job for the nation was made dramatically evident during the first Gulf War in 1991. Manufacturing workers in Eufaula, Alabama, were told of a shortage of rail launchers that attach the *Maverick* missile to various strike aircraft. They volunteered to work twelve hours a day without overtime pay for many weeks to double production output quantities. Management had to lock the doors on Saturday nights to prevent excess fatigue.

Locations offshore were also important in expanding our product lines. Facilities in Vancouver, Canada, were essential in winning *CAATS* (Canadian Air Traffic Control System). We also operated excellent plants in Mexico, Scotland, and England. These facilities were especially advantageous because of their very low labor costs, high product quality, or professional expertise.

Government Handicaps

With a noble objective to reduce costs through competitive pricing, the government began **second sourcing**. There were to be at least two manufacturing firms bidding to capture portions of each year's procurement. This worthy objective makes much sense in the usual business world of the United States. However, in the case of sophisticated devices bought in small quantities, this method was badly flawed. The "home" corporation had spent many years and its own funding and expertise in the advanced technology needed to create a unique machine. Such investments must be recaptured through production output earnings. The specialized manufacturing tooling and production personnel training also took time to mature and were quite costly. The procedure tried by the government was to select another company (not competitively), pay them to establish production implementation, and require the "mother" company to pass on detailed technical skills as well as coach the other company's manufacturing personnel. Annual bidding was then done for portions of the customer's small total quantity (limited by DoD budgets) for that year.

Differing portions of that quantity were then awarded to each contractor based on the price they had bid: the low bidder could capture up to 70 percent of that annual lot. The resulting number to be produced by each company was quite small. The manufacturing cost of these types of products is very closely dependent on production rates: the higher the number, the lower the resulting unit cost since the large implementation investment is divided by a larger quantity. Because of this, competitive unit prices were bid a little lower than previous experience from the mother corporation, and there was not nearly enough DoD cost savings to offset the manufacturing setup costs "given" to the second source. It would have been far smarter to double the rate at motherland, thus substantially reducing unit cost. Extensive government auditing of manufacturing cost charges, already actively in place, would be sufficient to detect any inappropriate expenditures by the "home" company. This would also complete the total quantity desired for deployment much earlier, and there would have been no wasted expense to implement and train the second source.

Setting Raytheon up to also produce *Phoenix* missiles was very foolish, costing the government several hundred million dollars more than if we had produced all the 6,200 eventually manufactured over a span of twenty years. Competition is always an excellent buyer's ploy, but in this case, sole source would have been far cheaper to the government. Additionally, continued second sourcing with the mandatory openness of design and manufacturing techniques on current programs eventually would make us all the same. How then could there be viable future competitions?

The DoD's production purchases of military hardware were normally only year to year, being constrained by the uncertainties of annual budget squabbles with Congress. I always thought this was crazy since the government would clearly gain a cost benefit if **multiyear** commitments were permitted. Mentioned in chapter 9 is an aborted effort we attempted to reduce the cost of *TOW* missile. These weapons were purchased by the USA in large quantities each year. (Perhaps this could have been an effective item for a competitive second manufacturer.) There was certainty that purchases would continue for many years in the future. Even informed Congressional members could see that. But unbending policy ground rules were prohibitive. Great savings can accrue by buying several years of components and placing them in inventory until needed for final assembly. As described later about *TOW*, a prime contractor trying this could be subject to financial penalties, even if the contractor took the risk of no further purchases of the final product or shared the benefits with the government.

We at Hughes were otherwise little affected by political intrigue, but throughout my career, the name of our founder, Howard Hughes, faced a **negative bias** in Washington. There was long-standing resentment about his many odd behavioral practices with the government, as well as his well-known inability to make rapid decisions. This inner feeling by politicians persisted well after his effective separation from the company operations in 1953. We had to overwhelm that bias by continually demonstrating excellent results in our military projects.

Woes of Being a Subcontractor or Supplier

It will be very frustrating when any R&D "baby" is not chosen for **in-house manufacturing**. The big earnings flow from large production volume. Maturing the design may have required internal assets, whether funds or skilled talents, if the costs exceeded the amount stated by the buyer in the initial agreement. Hughes perfected the advanced infrared seeker for the FIM-43 *Redeye*, a shoulder-fired surface-to-air missile. In spite of a very successful evaluation of our design, and enthusiastic approval by the USA customer, the prime contractor, General Dynamics Corporation, decided to produce the entire weapon and not subcontract the seeker to Hughes. They naturally owned the design (without reimbursing our

Redeye Launch
(© By Armémuseum [Swedish Army Museum])

self-funded shortfall and use of unique skills), but this applied a big negative for future relations between us. Over 85,000 were built. Rats! We lost a big lucrative business gain, in spite of hardware excellence!

In high-technology competitions, success usually results from becoming the first to discover a solution to a perplexing performance shortfall. Long-term business gains stem from such inventions, and key hardware elements of the design can be protected from competitor usage by federal patent protection. It is likely that people can dream up many innovative uses of this creation, increasing the company product lines, as can be done by others. Since 1960, when we made the laser practical, 55,000 U.S. patents have been awarded for different applications. These were so different from our patent's usage listing that we gained little compensation, and of course, the older seven-year limit had soon been reached (the protection span is now twenty years.)

In a radar case described in chapter 2, we finally perfected a device that solved a critical limitation of airborne radars. Such radars could not spot small aircraft while casting beams toward the Earth. Enormous reflections from the terrain overwhelm modest signal strength bouncing back from any tiny object. Military air defense and in-flight commercial airliners sorely needed what was termed **look-down** properties, whereby other craft flying beneath them could be located. The triumph became possible with our TWT, a frequency-stable power amplifier component.

With protective patents and quality production, Hughes became the world's largest supplier of TWTs. Patent coverage was extended by continuing design improvements. This market dominance was most rewarding. However, senior executives were often faced with a business dilemma: If there would be an upcoming radar system opportunity selling TWTs (the only device available to solve that look-down problem), our competitive advantage for the entire system might be jeopardized. Winning the big radar would gain far more revenue than selling individual components. In a number of cases, however, the government compelled us not to deny our opponents to use TWTs. Sometimes being a successful component supplier can hurt the parent corporation.

Undecipherable Losses

In 1984, the FAA (Federal Aviation Administration) issued contracts of $200 million each to teams led by IBM and Hughes to design and demonstrate the technologies for a completely **new air traffic control system** for the U.S. airspace. The feasibility project was called Advanced Automation System (AAS). After four years of evaluations, a brutal competition ensued for a multi-billion-dollar design contract, to be followed by long-term support and performance upgrades. This bonanza might be followed by an extension for international control networks. Our team included Sanders Associates, a unit of Lockheed, for controllers' equipment and displays; Unisys for dispersed computers and data communication devices; and a division of IBM (an interesting possible conflict of interest) for central computers.

Air Traffic Control Consoles

The IBM team won this glorious gem. We filed a formal protest asserting that the IBM proposal had not complied with stated bidding obligations in six major performance requirements. Astonishingly, our opponent's price also was higher than ours. After prolonged hearings, the contract award remained at IBM. It was speculated that much political pressure had been brought to bear and that the conservatives in the FAA feared more risk with that "startup" Hughes. That iconic IBM bore a fine reputation. The U.S. suffered from that choice: performance fell far short of desired levels, schedules were delayed for years, and costs ballooned. Would Hughes have done any better? That is moot. However, our company's track record on many other complex systems was excellent in the next decade. Included were air traffic control systems successes in Canada, Korea, Switzerland, Egypt, and Saudi Arabia.

ICBM defense is best done by destroying the missile during the boost phase or in the midcourse phase before it disperses the clusters of reentry vehicles with warheads and many deceptive objects. In mid-course, the hostile warhead carrier is traveling at 15,700 miles per hour above the atmosphere. If this opportunity to stop it is lost, it becomes mandatory to intercept and destroy all the reentry warheads before the nuclear blasts are triggered. To avoid wasteful intercepts, decoys must be identified and ignored.

Flight tests in the late 1980s confirmed the performance of the Hughes experimental *KKV* (Kinetic Kill Vehicle). It had solved many technical challenges: a seeker finding and tracking the hostile; distinguishing targets from background interference; having agile steering at closing speeds of Mach 10; and striking the most vulnerable part of the ICBM carrier bus to maximize destruction. The cylindrically shaped bird was eight inches in diameter and two feet long, with no aerodynamic wings or steering control vanes. Closing velocities would be 2 miles per second, enough for target destruction by kinetic energy, without an exploding warhead. It was boosted to 90-miles altitude from a ground base or ship and had an internal rocket motor for its final stage. Steering direction was governed by eight hydrazine gas nozzle thrusters on its periphery. It sported two "bug-eye" IR imaging sensors. These locked on to the target and analyzed its shape for precise steering. Final attack began at 60 miles; *KKV* travelled 30 miles to impact. Many test launches against ballistic missile reentry vehicles confirmed this intercept concept.

The USA set up a program called *THAAD* (Terminal High Altitude Area Defense) needing such a terminal intercept device. Lockheed won this very difficult competition. We were told that the USA's technical

THAAD Launch
(USA photograph)

evaluators favored our design, proven in many test shots. It appeared to us that Lockheed won on a political basis by promising to establish a new manufacturing plant in a state represented by a most influential Congressman. In the ensuing test program, Lockheed's design failed miserably, and the USA directed that the configuration be altered to mimic ours! Such are the ways of government competitions in which the loser's design is hijacked by the winner.

Our basic design concept was especially well demonstrated in 2008. Launched from a USN ship, a Raytheon *Standard* missile booster topped with a derivation of the Hughes KKV intercepted and destroyed a U.S. satellite. That target was prematurely decaying from its orbit and contained a large tank of unexpended toxic hydrazine fuel. To ensure vaporizing that chemical before it polluted at ground level, the missile software was preprogrammed to place the impact point directly on that fuel tank. Intercept occurred at 153 miles above the Pacific Ocean and the shot worked perfectly, much to the relief of many of the world's inhabitants.

Why Frequent Victories?

As you can see from all the preceding examples, competition for Pentagon programs can be won or lost by many interactive factors, not just the proposed price and delivery schedules. Hughes offers were always bolstered by our solid track record of meeting and exceeding difficult technology advancements. We relished winning 80 percent of all competitions that we entered. This enabled us to be foremost in achieving technology breakthroughs in every segment of the electromagnetic spectrum and to grow from 15,000 employees in 1952 to 85,000 in only twenty-five years. We had over thirty sites in California and operating facilities in thirteen states and four other countries. Our diverse customer array included fifty nations, and we were the leading satellite developer and producer for decades. We also were proud to become the largest non-government employer in both California and Arizona. How was this possible? Here are five reasons, some already stated in chapter 4.

First, most employees were imbued with the legacy of Howard Hughes: to strive to set new performance records, take risks to achieve technical breakthroughs, design for excellence, and seldom reveal corporate goals or our work to the public or to any competitors.

Second, we dedicated our efforts to the critical U.S. mission to survive the Cold War with the Soviet Union. This national goal, endorsed by the public, assured adequate funding and motivated all of us to invent devices far superior to those being developed in Eastern Europe. That "War" lasted almost fifty years without combat, and our nation prevailed; the Soviets experienced economic collapse. They could not afford to continue spending vast sums trying to catch up with our country's technical advances.

Third, we enjoyed the financial cocoon of private ownership. As a subsidiary of the Hughes Tool Company, and then HHMI (Howard Hughes Medical Institute), we were buffered from the public stockholder demands for profit growth in each succeeding fiscal quarter. We negotiated a single annual dividend payment to HHMI, usually a modest amount, so were thus able to liberally invest in internal research and development. Long-term objectives prevailed, a necessity for perfecting complex creations. This strategy fostered mutual trust and stability for our staff and resulted in rapid growth in our product lines and business base.

Fourth, we evolved a superb recruiting system to attract and capture the best talent in the United States. The DoD security classification clearance rules precluded us from hiring foreign professionals. Our location in Southern California was most helpful, with its mild weather, recreational wonders, and an ambience stimulating risk-taking and creativity. For many decades, staff members were attracted by educational opportunities, salaries higher than those of similar corporations, a semi-annual interactive performance review, and public recognition for their particular contributions. All these attractions yielded a high-talent staff of 22,000 engineers and scientists, including 4,000 PhDs. Our leaders continually promoted and convinced the staff that we together—functioning as a family and sharing technology expertise—could tackle complex problems to assure mission attainment.

Fifth, Hughes always placed a premium focus on assuring and sustaining customer satisfaction. This was done by endeavoring to understand their unique mannerisms and means of communication, responding rapidly to their concerns and desires for change, and maintaining amicable contact throughout the life of all programs.

As aptly stated by Ed Cobleigh:

> One of the key elements, in my opinion, for the successes of Hughes was the corporate culture that promoted having the non-engineering, non-manufacturing elements on board with overall corporate goals. I had only worked at two aerospace companies, Hughes and Piper Aircraft. I thought this optimum state of affairs was automatic. When we were bought by GM and later managed by Raytheon, I found out that this organizational bliss

was not a given. At GM, I saw the various divisions, Chevrolet, Buick, etc., working for their own successes, narrowly defined, and ignoring overall GM financial health. The road to the thirteenth floor in Detroit, such as sought by the manager of Chevrolet, was to maximize the sale of Chevys. If this cannibalized sales from Pontiac, well that was just too bad. I could bore you for hours with stories along this line. At Raytheon, we had a different problem. The various functions, Finance, Human Resources, Contracts, and especially the legal department, but certainly not Marketing, had their own criteria for success, which at a minimum were not plugged into overall corporate goals, and at worst were in conflict with those goals. It would take many moons to recount the dysfunction that resulted. I didn't know how good we had it at Hughes and still don't know how we did it.

9

KEEPING THE COFFERS FULL

MANAGING FINANCIAL STRUCTURES for prolonged periods can be a very difficult problem. There can be an almost unlimited set of daily expenses requiring cash flow, numerous large allotments at periodic intervals, and frequent surprise events making further demands for funding. These costs must be supported by income from current sales, existing contract payments, or perhaps by infusion of new investment offerings. All stockholders and capital investors will expect that total earnings exceed outgoing expenses by a sufficient amount to yield asset gains for them or the possibility of long-term business growth.

It is hoped that the contents of this book will inspire corporate leaders to place high priority on dedicating a significant proportion of earnings to **unwavering budgets for IR&D** (research and development funded from internal sources). That is the best and perhaps only way to keep the firm vibrant and successful. It is essential that senior executives continually persuade the board, the stockholders, and capital investors that this priority must be prominent, even when resulting in reduced profits in the short term.

Government stabilized **monetary systems** are so essential every day in most cultures. There are innumerable popular descriptors of currency "hardware." There are also electronic storage and transfer, value measurements that soon will replace all physical paper devices. Amusing U.S. monikers are "dough," "grubstake," "lettuce,"

Treasured Gold

"loot," "moolah," and "stash." Values of everything—whether hardware, services, or national and corporate viability—can be described by numbers in a monetary system. These began evolving about 5,000 BCE in the Middle East, using metal objects of various shapes to exchange in trade for essential and ornamental commodities. The first fully documented coinage was used in Ephesus, Ionia (now Turkey), about 650 BCE. The coins were composed of a natural amalgam of gold and silver. Today, there is still some direct bartering exchanges of trade-goods or services, but most nations have a formal monetary system, and there are dynamic international standards used to determine rates of exchange between different countries.

Assuring Financial Stability

All clusters of people trying to accomplish something new and useful need a **continuing source of funding**. It is possible to achieve good things using only "volunteer" labor, as was tried in the hippie communal gatherings in the 1960s. But most cultures desire and demand to live better lifestyles, with comfortable housing, safe sanitation, information and entertainment devices, adequate transportation, as well as good nourishment, clothing, and medical care. Creating and supplying these require the efforts of many; access and use of natural resources; and investments in land, buildings, and machinery.

There have been many attempts to find political ways for everyone to receive a fair share of what's available, but so far the only system that has been successful for long periods is the capitalist society that does not embody too many negatives. It has many flaws, including excessive use of natural resources (it's reassuring that excellent progress has been made in renewable energy and material recycling); blatant greed by the wealthy and investment speculators (continual government pressures have been somewhat successful); and inappropriate nepotism in passing wealth and leadership reins to successive generations. There has been growing public support for rewards going to the best talents and those who demonstrate positive results. There have been valid outcries about excess remuneration given to corporate executives. However, it seems odd there is little complaint and even envy regarding the bizarre amounts paid to movie stars, rap artists, and professional athletes. Credible critics of this economic approach point to concerns that moral standards will decline if there are substantial discrepancies between the wealthy and those not well off. This is a very appropriate concern, but it encourages unworkable proposed solutions, such as a guaranteed income without any active work or contribution to society. There must be some room for personal responsibility and motivation to grow and improve one's living conditions. History has shown that a stable monetary system with opportunities to increase personal assets has been the best path for a culture. No system is perfect, so there must be continual adjustment to counter adverse effects to the overall society. But, no matter what is done, society must support and motivate innovation.

In the capitalist structure, corporations require a firm financial base, traditionally termed its "capital assets." Such funds are used to acquire land and facilities. Funds must be sufficient for initial cash flow for purchasing materials and services, cash to pay employees, investments for providing future employee benefits, and funds to repay borrowings. All these latter costs require continuing sources of revenue. The long-term success of a firm is strongly influenced by its wisdom in managing capital expenditures.

There is much criticism of those organizations driven by the profit motive. Such profits can be used for directly rewarding founders, reducing debt, adding to corporate asset investments, and paying dividends to stockholders. Increasing profit levels can also stimulate improved prices for the publicly traded stock holdings. However, it is intended herein to persuade corporate decision makers that the optimum earnings usage is to allocate a large proportion to internal research and development. Benefits of this practice will be enhancement of competitive posture and, likely, a significant expansion in size and market share.

Where Does the Dough Come From?

To thrive and prosper, any corporation must seek and **find reliable founts** of money. Sources will vary as the firm matures, changes its objectives, incurs difficulties, or somehow becomes independent of external help. Start-up entrepreneurs may be able to use their own assets to form a company financial base, or solicit investments from venture capitalists. That has the negative of shared ownership, which is even more weighty for the next expansion step: offering stock certificates to public investors. Each time more shares are introduced, there is further dilution of corporate ownership. The remedial strategy by the company founders is to maintain control by retaining the voting majority of total shares. Some stock types are issued with no ownership rights, only a proportionate sharing of dividends. In some start-ups or expanding firms, the investment amounts to establish new facilities and equipment can be lessened by the customer providing the resources. In 1951, there was critical need for large inventories of high-performance guided missiles: the USAF opted to pay for construction of the facilities and manufacturing tools placed on Hughes land acreage for a new plant in Tucson, Arizona. Henceforth, Hughes paid monthly leases for use of those portions of the plant.

Large amounts of funds can be garnered from borrowing: mortgages on facilities, bank loans, government advances, or sale of bonds with corporate assets as collateral. These tools can be especially helpful when adding facilities or equipment but are ill advised as a means of overcoming cash flow shortfall for daily operations, unless only temporary.

Armored Money Transportation

The **primary source** for funds for daily operations, enabling profit, rewarding investors, and achieving growth comes from customers purchasing your products or services. The diverse markets range from consumers, other corporations, overseas nations, or elements of the government. For aerospace firms, the latter include DoD, CIA, NRO (National Reconnaissance Office), NASA, and the FAA. Much of this book is about finding ways to assure strong and continuing investments in research and development. The government for years has been a solid source of support for leading the nation ever upward in advanced technology. Staying ahead in the global economy race is vital. However, there is growing public pressure to reduce budgets for exploratory research on projects that have no apparent purpose. A good example is the recent bureaucratic proposal to reduce the stature and budget of DARPA (Defense Advanced Research Projects Agency), a free-wheeling agency empowered to fund inventive minds on what may initially seem to be nutty concepts. This free-thinking attitude yielded many breakthrough technologies that grew into significant advances in our defense posture during the Cold

War. Those new discoveries, as usual, soon morphed into never-before-seen devices that greatly benefit mankind. DARPA's modest budget was easy to uphold in Congress with that Cold War threat looming, but not so today.

A questionable way to garner government funding for R&D is called "bootlegging." If a defense product is in an on-going manufacturing contract, it is appropriate to have supporting engineers in the factory to solve occasional design configuration difficulties that may arise. It may be appropriate to have a number of such engineers on site to produce rapid solutions to avoid costly production delays. If there is no problem that week, these professionals can make good use of their skills toward designing something new or finding more efficient production procedures. Adequate production support must be funded by the customer, so the new work may appear on audit as "free" bootlegging. However, if the new work improves the product, as it usually does, all parties benefit.

Sometimes the inverse bootlegging can happen, wherein the customer product is improved by an overzealous design engineer voluntarily adding a new performance level beyond the contract requirement, without a negotiated payment adjustment. Once at Hughes this had occurred on a major radar program being developed on a subcontract with McDonnell Douglas. Hearing of this during a large meeting of our staffers, I inquired of that superlative design leader, "Have you gotten your paycheck this week?" After he said that he had, I asked, "Are you sure you will next week?" Needless to say, the contract had a corrective codicil a few days later.

A boost in cash for use in internal IR&D projects can come from a bonus awarded for contract execution excellence, whether for early delivery, hardware performance exceeding the agreed goal, or promised lower manufacturing costs. Such incentives are quite common in both government and commercial contracts since all participants gain the benefits.

A real continuing income surge can arise from selling to other firms the ability to use in their products a feature that your company had patented. Most firms own patent rights for anything created on their premises or done with internal or customer funding. At a minimum, the individual creator is named in the document, a considerable kudo adding to professional prestige. Hughes prospered for many years as the leading supplier of TWTs and lasers (see chapters 2 and 11). Both were protected by exclusive patent holdings. We financially benefited from both hardware sales and patent usage fees.

Sometimes lucrative results can arise from settlements of lawsuits filed against another firm for **patent infringement**. Here are two examples of pirating our inventions that resulted in richly rewarding settlements.

A 1973 patent was issued for satellite attitude stabilization control (ten years after Don Williams devised the concept!). The work had been performed with company funds, but the government allowed other contractors to use this gem without recompense to us. We filed a lawsuit against NASA and the manufacturers of 108 satellites. Hearings determined that ten birds did not infringe. In 1988, Philco-Ford settled thirteen infringements out of court for $45 million. The remaining eighty-five infringements committed by the government agency were finally resolved in our favor after an incredible twenty-six years, for $154 million. It took a mighty long time to harvest that crop!

The second example was not that good. In the 1970s, we filed against Intel Corporation for $200 million. Dr. Robert Bowers of our Research Labs had perfected a new process for making semiconductor circuit devices: clusters of hundreds of microelectronic elements were created by etching and altering the chemistry of silicon wafers and assuring far better placement and high quality. This process is now used in virtually every electronics device throughout the world. Unfortunately, we lost in court because of a single juror's negative vote. However, smaller claims on the same patents were negotiated out-of-court with several Japanese firms yielding almost $20 million. In spite of the opinion of that one strange juror, the Japanese believed we had a very strong position.

Many companies derive large and regular income from product sales to the public, especially to vast numbers of insatiable consumers. We live in a culture of lavish resource usage to pamper personal comfort. In many cases, this is the very reason for thousands of firms to exist. Essential for survival and gain here is continuous attention to unit cost, quality, and market pricing, as discussed below.

Using Your Assets Wisely

A very effective way to **maintain balance** between investment commitments and magnitude of technical risk is illustrated by the strategy used to achieve the very successful *DirecTV* product, providing television service directly to individual consumers. The concept was to plan development in successive phases and coax government and network suppliers to fund large portions for each step. It began in the 1970s, and it took twenty years to mature into triumph in 1994. Planning was heavily influenced by the need and source of capital. Four business funding project types were undertaken in sequence, with magnificent results. By 2010, there were nineteen million customers generating more than $20 billion in annual revenue. Most rewarding and astonishing was the AT&T purchase of *DirecTV* for $48.5 billion in 2014—an astonishing payoff for our $200 million investment. Here are the steps taken to capture the bountiful market:

- Phase 1: Design and manufacture satellites with all funding coming from the customer. The user operated the network. The Communications Satellite Act of 1962 opened the skies to commercialization and established an administration agency called Comsat (Communications Satellite Corporation). Its mission was to lead the development of a worldwide communications network. In 1964, it joined with seventeen other countries to form *Intelsat*, a consortium of communications companies. Intelsat would own and operate the global satellite system. We won the initial competition and supplied *Intelsat-1 (Early Bird)*, the first geostationary global commercial communications satellite. In the following years, many higher performance birds were developed and orbited for operations by many nations.

- Phase 2: Develop and build satellites using company funding and operate the network with costs paid by a single user. The 1978 USN's *LEASAT* procurement was awarded to us. Hughes funded the development and production, but money returns were guaranteed by a long-term operating lease. Orbit injection was from the Space Shuttle, necessitating a significant spacecraft redesign. To keep the network functioning, we staffed the orbiting controls and Earth-based segments, thereby establishing our own credentials as both an equipment supplier and a system-wide operator.

- Phase 3: Design and build an entire orbiting and ground station system, using company funds; own it; operate it, and sell many separate relay links to companies in the business of furnishing arrays of TV channels to consumers. An expanded "open skies" policy by the United States in the 1970s allowed ownership of communications satellite systems by any outfit approved by the FCC. We viewed that this business setup could be quite lucrative and formed a subsidiary, HCI (Hughes Communications, Inc.), to provide the ground stations, sell usage rights for each transponder, and be the service provider to those users. This venture was quite risky, with unknowns: sufficiency of marketing of networks to new customers; keeping them happy; the life-span of the satellite; adequacy of relay power levels; equipment reliability; and coping long-term with operating costs and currency inflation. Fortunately, we found that the cost to get each relay transponder fully functioning was $3.5 million and resulted in revenues of $16 million. Pretty nice! Customers were eight major TV network firms. Mature versions of our Galaxy, with ground stations around the world, relayed eighteen transponders, each with the capacity for five to seven TV channels and 1,000 voice and data linkups.

- Phase 4: Enable direct sales to individual consumers. Our culminating step was to provide services directly to individual businesses and homes, a sustainable market to millions of customers. One great concern was that this would compete directly with cable distribution companies, our best clients in phase 3. It would also need a very different and large marketing staff. What emerged was *DirecTV*, using our new *HS-601* satellite. The system's technology base was well in hand: high-power broadcasting over a large area into small antenna dishes with stationary pointing angles. Each customer could continually capture up to 250 TV channels.

Cost Corrals

Survival, growth, and prosperity of businesses are directly tied to their cost levels. Small percentage changes can create an enormous effect on overall corporate results. Constant attention is needed to control any deviations. Everyone knows that in the consumer market, aggressive pricing tactics (described later in this chapter) yield boom or bust sales. For resultant financial success, the underlying cost structure can have an adverse or a sunny outcome. Some expense can be adjusted with effort, but in a large and highly equipped outfit some costs are relatively fixed for uncomfortably long periods. Let's briefly list these by type, using the traditional accounting jargon, and follow that with some hints on how to maneuver some of them.

There are relatively malleable expenses including salaries and bonuses, travel and incidentals, quarterly dividends, set-asides for coping with potential risk, some tax levels, and IR&D budgets. Others are leases and rentals, insurance for facilities and equipment, repair and upgrade, loan principal reduction, and outside investment of surplus earnings. Those that are negotiable include subcontract development and production, component and material supplies, product delivery transportation, wholesaler and store charges, employee benefits for medical, and retirement. Be aware that once the latter two amounts have been set, there will be little opportunity to decrease them in the future. There are other expenses that are little subject to favorable change, including taxes of many kinds, import-export fees, interest on existing loans, patent rights payments, and building depreciation.

To get a handle on some of these, consider these ideas: Assign your best talent on the job, whether initial development or production, and thereby relish fine productivity and quality. Output will emerge faster and there will be few flaws to correct before delivery. Make sure the entire workforce understands and personally adopts the firm's goals and objectives. Also inspire all on every team to excel. Provide the best equipment and tools to assist the workforce. Arrange a cooperative bond and open communication between engineering and manufacturing teams, overlapping the development and testing phases. Encourage that factory planning is done with process time-sequence layouts and employee placement and training. Foster rapport with all suppliers: we're in this venture together.

Manipulating salaries and wages may not be very easy. As mentioned previously, these are critical to employee morale. Once a precedent has been set within an industry or in a particular firm, any attempt to decrease that standard will fail. There will be an instant dive in staff morale and satisfaction. Also, government rules impose minimum wage levels, competitors may pirate the best parts of your flock, and attempts to hire new staff will be highly influenced by the starting wage and benefits offered. There are times when talent, especially high-capability professionals, is in short supply. Employees may choose to hopscotch between different companies, benefiting from negotiating successively increasing incentive offers.

Several tactics can be helpful to control salary and benefit expenses. Optimal matching of talent with the job to be done will increase productivity and reduce error, thus getting things accomplished efficiently at less cost. Scientific scheduling of the collective tasks can be sequenced to accomplish all with a smaller staff rotating between different programs. Some hands-on work can be done more quickly and cheaply by automated equipment. An adept subcontractor may get the work done at lesser costs than the in-house facility. Hardware redesign can greatly reduce material, tooling, and labor expenditures.

Many U.S. firms have greatly benefited from **outsourcing labor**, particularly manufacturing, to other nations. Lesser labor, overhead, and taxation charges may enable your product to appear in the marketplace at two-thirds the price if it had been locally produced. This has become a political dispute; critics claim it reduces domestic employment and avoids U.S. taxes. However, consider that the item will be more affordable to consumers, more successful in overseas markets, and will give the supplier country better local employment and ability to purchase other U.S. products. Hughes setup four *maquiladoras*, factories in Mexico run by a foreign country, to produce at high rates many types of sophisticated electronic assemblies for installation into our systems. We found that dedicated and highly motivated staffs produced output of the highest quality and lowest cost possible. All those outfits were judged by the Mexicans as the finest workplaces in their cities.

Another effective way to reduce unit cost is **high rate** production. This can significantly lessen the impact of overhead (facility depreciation, plant maintenance, loan interest, etc.) charges. Those are relatively fixed in a given time period. Part of unit price is used to pay for them. When a higher volume of product is produced in that time period, the total overhead is divided into many pieces, thus reducing unit cost. High rates can be scheduled by compressing that year output demand into a six-month span. The workforce and the facility may be employed for other projects (if you have them) or leased to other users.

Multiyear Denied

Manufacturing cost reduction can also result from **multiyear contracting**. Such agreements can be reached with commercial firms but are difficult in government procurements that are limited by budgets needing annual Congressional approval. If arranged, these can enable the above described compressed schedules strategy, especially for suppliers of small components. They also provide more security from inflation and pricing escalation by suppliers.

We attempted this cost-savings method, which triggered legal action in Washington, with accusations that Hughes had deliberately mischarged the government for one year of *TOW* (Tube-launched, Optically-tracked, Wire-guided missile) production. For years, it had been in high rate manufacturing (almost one million have been fielded by the USA, USMC, and forty-nine other friendly nations). Its guidance required command linkage from the launcher directly to the bird in flight using an internally spooled wire pair trailing up to 2½ miles until target impact. The lightweight wire had to be very strong, flexible for tight spooling and rapid spin-off, and not be likely to tangle during in-flight dispersal. We had teamed with US Steel to design and manufacture this very unique genus

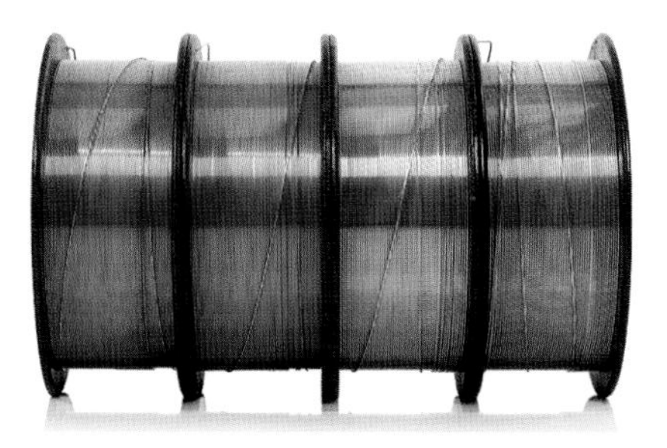

Spooled Wire

of hardware. To take advantage of lower unit cost from higher rate production, one of our aggressive managers negotiated with US Steel to buy in a single lot the inventory we would use for the next three years, using our own funds. This was risky to us, with future Congressional budget authorization unknown. There was some likelihood that we would be stuck with a two-year supply of expensive spooled wire with no alternative use. Our enterprising *TOW* staffer thought it appropriate in pricing the next year's quantity to include the spooled wire cost at the price US Steel would have charged for a single year quantity rather than the lower amount stemming from our multiyear purchase. This is clearly a no-no in procurement regulations since a prime contractor can charge no more for a subcontracted item than it had actually paid. Our internal auditors had not caught this error, thus the government lawsuit against Hughes. That was settled after a long period of negotiation, with some penalty amount added. Here is another case to be examined for potential government cost savings: if a prime contractor takes on the risk of receiving no succeeding orders by buying larger quantity lots of some component types, why not make it legally possible to at least share the resulting significant total cost savings?

Here is a positive experience by Ed Cobleigh:

> Another way to benefit from **high rate** manufacturing is to apply commercial components in military hardware. On the *Tomahawk* winner-take-all competition, we discovered the chip that controls the little jet engine cost $1,800 each. We called some guys at our owners, GM, and found out the engine controller for the Saturn automobile did the same job. Will it qualify to MilSpec (military specifications) standards? It was better. The under-hood environment of a car is worse, except for altitude, than in a cruise missile. We had to make two circuit changes and qualify the chip at altitude, and we

were in business. GM's chip foundry asked us how many we needed. The answer was maybe one hundred a year for ten years, bought a year at a time. That got a lot of laughs from GM. The plant manager told us he would have the folks come in on Saturday morning and bang out all one thousand chips at once, and if the out-year orders for *Tomahawks* didn't come in, then just throw the chips away. What made this a viable plan was the cost of the modified chip, $43.00.

Pricing Maneuvering

Outputs of most business organizations rely on **on-going sales** for their long-term livelihood. Regardless of what the product is, its success in the market will be much dependent on its price. A thoughtful and mobile strategy must always be active to determine and respond to market reactions. If your product performs better than any competitor, things are a bit easier. If not, a small alteration in the current price can make an enormous change in sales volume. Actions by aggressive competitors can wield much damage, perhaps without remedy. It would be nice if you could accurately compute your cost to produce and distribute the item, add a reasonable profit margin, and press on to the marketplace. That seldom occurs. There also can be negative impact on your total volume caused by distributors, such as department stores, that will set prices using their own marketing tactics.

Many people think that it would be most enjoyable for consumers to purchase products only from the efforts of **nonprofit** organizations. But those dreamers should consider that such outfits are deeply dependent on donors to provide the facilities and equipment, volunteer workers, tax relief from governments, and "free" distribution methods. Particularly acute is the willingness of nice folks to repeatedly send large sums to this particular "nonprofit" outfit. A drying up of donations may be fatal. History also shows that in such organizations there is poor incentive to productivity or striving for excellence. These factors can mean that if it is a service being provided, this operating method can be viable. However, for producing hardware in large quantities, the work done by competitive profit-oriented firms is far better for creating devices for consumers. For the buyer, profit to the producer is more than offset by lower unit production costs that these firms must always pursue.

Other than striving to reduce cost and setting minimum profit levels, here are a few hints to think about in determining the sales price of an item for the consumer marketplace. Start with **low prices**, even resulting in temporary losses, to stir interest. Follow this with gradual increases as acceptance blossoms. This tactic can also be used to squeeze out a competitor, as happened to Hughes with the digital watch, described in chapter 11. Another sales aid is to grow a noteworthy **brand label** for the product. Memorable examples are Nike, Michelin, GE, and Hart Schaffner Marx. The idea is to implant in customers the belief that "If it comes from them, it must be the best," based on the track record of previous products with that logo. The fashion world enjoys almost unlimited sale prices. In many cases, the higher the price (regardless of its inherent value), the more it is desired by some fashion-conscious customers. Names such as Cartier, Gucci, and Louis Vuitton are to be envied. A breakthrough **model appearance** can also enliven sales, such as the 1953 Chevrolet Corvette, the 1965 Ford Mustang, and the 2001 Chrysler PT Cruiser. Many companies have tried to gain market share by having the product personally **endorsed** by a celebrity. This began with Babe Ruth in 1930. Dominant recently has been Tiger Woods. This author feels immune to such come-ons, but a Harvard study showed that, on average, sales increased by 4 percent following such endorsements. This seems puny, but can add up to billions for some giant firm. Another clever plan is to place your product at a noted **distribution outlet** with a magic reputation like Harrod's, Neiman Marcus, Tiffany & Co, and Sachs Fifth Avenue. It can also be very beneficial to market in states that do not add sales taxes to all transactions.

A major pricing goal is to assure enough revenue from total sales to at least restore the investment required to create this new product. This can be exceedingly difficult for **pharmaceutical developers**. It may take many years of laboratory work, costing millions of internal research funds to chemically define a new drug. To protect this design, a patent application must be filed, which for these products will have a twenty-year protective patent span. Before the drug can be released to consumers, however, the government must approve public use of it. This procedure involves extensive safety and effectiveness verification tests lasting as long as ten years. Now, off to the market, realizing that only half the patent protection period will be available to recoup the costly investment. Other companies will then mimic the design and market a generic version. Those companies can set a low unit price since no IR&D recapture is needed. Guess what emerges? The originator charges very high prices for ten years, suffering raucous public complaint for gouging "to grab excessive profits." Ain't fair!

Prices for sales to commercial corporations may dominantly be based on your firm's historic reputation for output quality. What customer firms desire from you is equipment that will enhance their own business

Marketing Appeal

performance. If new machinery is needed, customers will likely contract for the development of something defined to fulfill a specific need. In such cases, pricing will be the result of extensive negotiations, proof of cost forecasts, permission to monitor during the contract span, and potential sharing of risk. You can best overcome competitors by being more adept at accomplishing the job based on your reputation for excellence, communication ability, and integrity.

Programs for the **government** bear many roadblocks to setting a reasonable price. Competitive offers can be made at the discretion of each seller making a bid. There is little maneuverability after winning since all expenses charged will be audited and possibly disallowed. All physical activities will be under continual surveillance, sometimes with obsolete compliance rulings forcefully applied. These federal safeguarding provisions should be lauded, but often cause costly delays, unnecessary retries, and often some acrimony. Contractors can be at great risk if technical difficulties occur, as often happens, with resultant costs not being borne by the customer. Fortunately, design alterations desired by either party can be successfully negotiated at fair prices. The previous chapter includes some hints on how to prevail in competitive struggles for new government-funded projects. The beauty of seeking business with the U.S. is that it is an opportunity to serve the nation and also establish a strong foundation of research and development funding. Technical advances achieved in such work for the government can be applied in future commercial endeavors to create new products. A premier example is Boeing's morphing its success on the USAF-funded KC-135 Stratotanker into a four jet engine airliner, the lucrative world-leading varieties of birds designated with numbers in the 700 series.

Sealing the Deal

Most consumer goods are bought from a retail supplier, such as a department store, at a stated price and for immediate availability. Some clever buyers can negotiate a price adjustment, as a long-term customer or personal friend, or with the promise of future purchases. For commercial or government deals, long-term arrangements are usually needed and desired between supplier and customer. For these, it is necessary to find an optimum way to record the agreement. The content should be clearly stated to avoid future misunderstandings and give lasting reassurance that the mutual goals will be reached. The document is called a **contract**. It defines an objective, whether a piece of hardware or a service, the agreed price arrangements, and a schedule with measurable milestones. In cases where something new is going to be created, performance levels can be stated as a range of values, with minimum acceptable and higher ranges as objectives. Incentives can be offered as bonuses for attainment of superior results, as verified by controlled testing. Similarly, price adjustments may be included as bonuses or penalties for final cost and schedule outcomes. The total text must be legally binding during the agreement's duration, and both sides should engage legal advice for protection of their rights.

A Firm Grasp of Trust

Optimum agreements can be concluded when the two parties know and trust each other. Suppliers with a fine track record can set up flexible arrangements to continually alter the best path to the buyer's goal. There must be adequate margins of time and expense to cope with inevitable surprises that crop up when the "unknown" is being sought. Contracts between experienced participants have several marvelous foundations: a

handshake agreement, denoting mutual trust and confidence; a sharing of mutual objectives; and the imposition of few complex procedures or difficult barriers. The best government contracting methods I experienced over many years, beneficial to both supplier and customer, were deals for programs conducted in extremely high security. These were from DoD, CIA, NRO, as well as in subcontracts with other prime aerospace firms. Two illustrations follow. Perhaps the U.S. economy could improve if these methods were frequently emulated in future commercial and government contracting.

Joyous Black Contracting

The astounding SR-71 and *YF-12 Blackbird* (flying over three times the speed of sound, and spanning to Europe and back) projects provide a vivid snapshot of an aerospace creation of great significance with breakthrough technologies and ideal interaction between the government procurement agency and two excellent contractor corporations: Lockheed Skunk Works and Hughes. Our design required revolutionary "look-down-shoot-down" interceptor capability. This meant that the radar must detect and track targets, even if they appear below the *YF-12*, and a long-range guided missile that will strike hostiles at any altitude. More details will be found in chapter 12. The black **procurement deals** were very efficient, relying on mutual trust between customer and providers. Contracts were simple, stating end objective; minimal procedural

SR-71 *Blackbird*

dictates; rapid decision-making; shared financial risk; as well as only essential and brief paperwork reporting. It could not have been more ideal, and that mutual trust achieved a masterful success in a short time at the least cost. That has been lost in the last decades and should be reexamined for use in future development planning.

The Hughes portion of the job was the electronic weapon control system, with its unique radar, and the long-range guided missile. We were the industry leader in these types of systems, and much of our new design was an outgrowth of the configuration to be used in the F-108 *Rapier* that the USAF had recently cancelled (details are in chapter 12). Such blessed projects permitted selection of the best talent (enabling only a small staff), the absence of many unnecessary procedural constraints, few demands for multiple-level decision approvals, avoidance of cumbersome outmoded specifications, minimal government technical oversight. A vivid example of quick decision resolution was my experience with the Skunk Works. On my first day on the *YF-12* project, I asked that the aircraft be lengthened by one foot to provide enough installation space. Within eight hours, we received concurrence, and suffered no delay in proceeding on our designs.

Appropriately, close monitoring of budget expenditures and legal matters was effected, and comprehensive hardware evaluation testing was performed jointly by contractor and customer. These contracting methods allowed the staff to exercise free and creative thinking and quickly reach the end objective. The fabled Lockheed Skunk Works team, led by the venerable Kelly Johnson, was famous for creating the F-104 *Starfighter*, U-2 *Dragon Lady*, SR-71 *Blackbird*, and F-117 *Nighthawk* in very short times and resulted in superb performance. Critical to those successes was a sense of trust by the U.S. procuring agency, and its acceptance of risk-taking for rapid development progress. I have heard these four contracts were only brief statements in a few pages: "Here is what we want as an end item—you select the technology and proceed in an efficient manner that works best for your organization." Too bad other DoD projects do not allow performing with this freedom to create new things, especially when a reliable contractor has been chosen for the task! Many agencies fear that something terribly wrong will occur unless the contractor is closely controlled. Perhaps they should recognize that if the contractor performs poorly, most future customers will go elsewhere, and the firm will be doomed. Performing with excellence in vital national missions is a most energizing motivator to do things right.

The glories of black in those days, I understand, are not in vogue today for crucially vital secret projects. We are back to paperwork demands, excessive technical oversight, and tedious delays from multi-level approvals. Gone is the customer approach that said: "Here's our objective; proceed in the most creative and efficient way." In the good-ol' days, there was no delay to obtain design approval from many uninvolved people and agencies, and no papers to be filled in triplicate. When top-notch professionals were assigned under the cherished former technique, they got the job done cheaper and faster. I never could understand why most military development programs, even at lower security classification, were so cluttered by unnecessary monitoring and mountainous documentation. As in every tough job, things certainly can sometimes go wrong in fast-track developments. But adding many observers who cannot fix the technical problem adds cost and increases the time for correcting fumbles. Performance stumbles usually happen only 5 percent of the time and are quickly remedied, assuring continuing trust between contractor and customer.

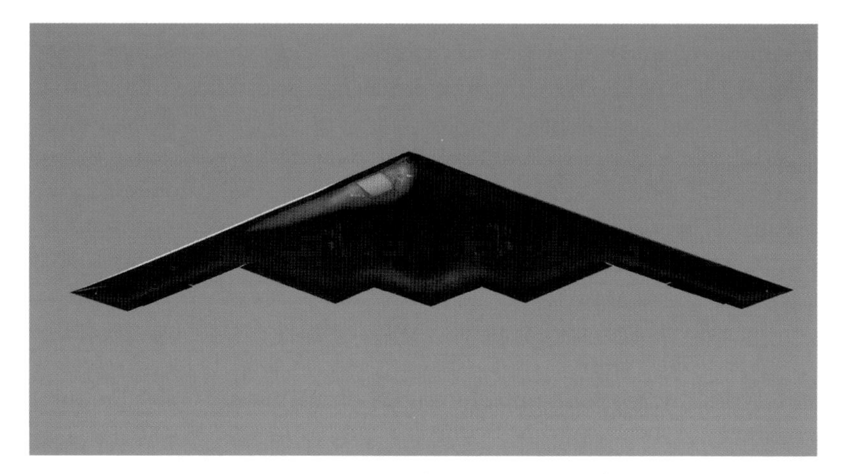

B-2 *Spirit*

The USAF program to develop the B-2 *Spirit* and its radar was another exemplar of the beauty of black contracting. The bomber vehicle and all of its equipment were intended to function in combat in a manner that would be undetectable by the hostile. The word **stealth** simply conveys that cardinal requirement. The principal cost stumbling block on B-2, however, was Northrop Grumman's and our ability to initially meet the exceedingly rigorous security rules required at all times. As you will read next, we actually helped Northrop install internally adequate security-control procedures after we had incorporated our own improvements. That demonstrated that we jointly formed a real team, even though we were only a subordinate subcontractor.

Hughes was chosen by Northrop as the radar creator. Once again, **mutual trust** and shared values allowed us both to reach the ultimate outcome for the USAF. Our stature included three experimental stealth radars and a fine long-term relationship with Northrop. Those successes were the JP-4 missile for the flying wing *Bat*, the radar for the F-89 *Scorpion*, the highly secret *Snark* cruise missile guidance, and the *Tacit Blue* radar. Stealth features are particularly difficult in designing the bomber's navigating, ground-mapping, target finding system. To do their job, such devices must emit electromagnetic energy and measure reflections from the area of interest. Such emissions may be vulnerable to detection by defense receivers. Instantaneous radical switching of frequencies and power levels can either provide confusion or be almost invisible because of undetectable signals being emitted.

Our initial contract value was $145 million cost-plus-fixed-fee and was based on a mutual assumption by the USAF, Northrop, and Hughes negotiators that existing security facilities and procedures would qualify, since both companies for twenty years had satisfactorily performed many other black programs. However, the new crushingly severe restrictions—probably necessary because of recent Soviet spying—were significantly tighter than anyone expected. Hughes flunked the intensive compliance survey, and all our operations were suspended until substantial facility reconstruction was completed and procedural doctrines were sufficiently upgraded. This cost an astounding $60 million above the basic deal. Fortunately, the government absorbed the cost because all the required improvements came in the form of a new command in direction from the USAF security organization. We completed remedial work within a few months and passed the government compliance inspection with flying colors and many compliments.

Northrop also failed its security exam, more extensively than we had. To solve this, and to alleviate manpower shortages, Northrop brought aboard many experienced people from Rockwell International (which had successfully developed the B-1 *Lancer* bomber) and retired USAF officers. This cluster of talent was jokingly known as North-Rock-Force. It is of interest that Northrop borrowed the new Hughes-developed procedure documentation, viewed by the USAF as an ideal model for compliance.

In spite of these security setbacks, plus some fundamental design difficulties, to the astonishment of both Northrop and the USAF, the first radar flight test in 1984 aboard the KC-135 *Stratotanker* flight test bed made recognizable high resolution SAR (synthetic aperture radar) ground map images. Each additional complex radar operating mode added into the equipment also performed successfully in its first flight. Such an achievement was a first in the history of radar full-scale development flight testing.

The *YF-12* and B-2 project successes are fine examples of the advantages of using the original black program contracting rules. National goals were met without intelligence leaks to potential enemies. The final configurations, with unusual performance effectively proven in numerous flight tests, were achieved at greatly lower cost and far quicker than by using traditional methods. To my knowledge there were no "crooked" activities, no profiteering, and no avoidance of excellence. Naysayers, demanding rigorous government control of "greedy" contractors, must take little solace for lack of evidence that few fears of corruption have been proven.

10

STRESS RELIEF

A PERSON IN ANY working environment must find several outlets providing stress relief. Daily occupational pressures can be of varying intensity and can stem from excess tedium, disappointment in job assignment, adverse reaction to other workers, poor supervision, concerns for the project's outcome, or the future of one's career. Stress mitigation can come in many forms; most successful are those that match a person's interests and capabilities. Results will depend on frequency of opportunity, willingness to train or practice, and one's enthusiasm to participate. Many activities require physical demands, so success and interest may change with one's age. Some can be done solo, but most require other participants. The latter provide stimulating opportunities to discuss and soothe one's inner feelings, perhaps even alter misplaced attitudes. A variety mixture with frequent switches between diversions usually is most effective to refresh and stabilize tranquility. It is essential to capitalize on leisure time. What follows is based on personal experiences over a forty-year professional life. These may give the reader some possible choices to make that will match their need and emotional inclination.

Leisure time must be provided by the employer. During the last half of the twentieth century, the work days in the United States lasted eight hours, weekends and holidays were at one's leisure, and annual paid vacations were granted. Vacations are often two weeks, growing to three, four, or more each year, based on an employee's longevity at that company. In some cases, overly dedicated employees may skip that annual benefit. The company will gain from the extra work performed. However, in the long run, this will be offset by that staff member being jaundiced, uncreative, and cranky. There have been many studies indicating that regular vacation usage benefits both employee and employer with increased productivity, crisp mental alertness, and restored physical health. One corporate enforcement tool is to allow the vacation salary payment pledge to expire after a two- or three-year accrual. One senior Hughes executive had accrued over five years and appealed that forfeiture to the Board for relief. A small compromise was negotiated.

The duration and time per day of a U.S. workweek in technology organizations today varies widely. Managers and coaches establish objectives for task completion, and the staff can personally fit their work times in order to meet the scheduled deadlines and assure project quality. This approach is helpful to the staff: each can adjust personal events, stress levels, and commuting arrangements. The firm gains by setting output results, rather than hours accrued, as the priority.

Enthusiastic Doodling

One simple relief from tedium is to casually doodle sketches of any type while sitting in exceedingly long meetings. It is somewhat risky since the activity is likely to insult the other participants. The doodling person appears disinterested and to be not paying attention. However, most agile minds can allow physical motion at the same time as focusing on a presentation. Folklore tales about Napoleon Bonaparte say he was able to do seven different functions at the same time without noticeable negative results. The boss when doodling should be very discreet and may also be shielded by being the senior person present (an unfair advantage!). That, hopefully, will not motivate others to doodle also. Many doodlers actually experience better absorption of the information being discussed. If one's superiors are present, it would be wise to avoid this habit. The author's associate Eileen Jennette prepared a framed set of many of mine as "masterpieces of art." These were displayed at a big gathering at a celebratory dinner. Those images revealed a background in high-school draftsmanship rather than classic painting on canvas. Eileen also was very helpful in delivering unsolicited health advice after secretly counting the number

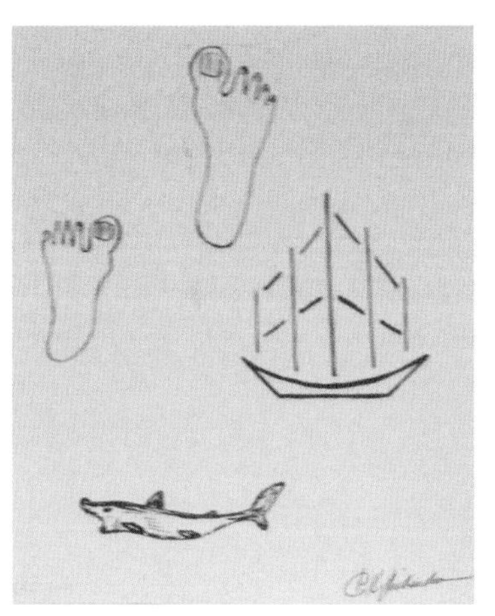

Priceless Doodles
(personal files)

of cups of coffee consumed. Upon measuring a ghastly twenty-seven per day, she recommended switching to decaf. A good thing, and it did not reduce the stress relief rendered by many java breaks.

Family Joys

A valuable aid to enjoyment of life is to be part of a dynamic and growing family. Finding, courting, and marrying can be fun, challenging, and mystifying. The long-term outcome must be resilient and flexible to adapt to changing times and personal attitudes. The correct choice of a mate can be quite difficult since emotions can be dominant and any individual's future behavior can be unpredictable. But the adventure can be very rewarding. One's spouse provides a responsive ear to troubled and stressful thoughts, may be a balancing factor in points of view, and can be a fine companion, rendering advice and sharing life experiences. Some companions can inspire excellence in personal behavior and professional accomplishments. From time to time, this relationship may actually add pressure to work stresses. But on balance, it is a considerable boon to one's lifetime existence, and far better than frequent bouts of loneliness.

A fulfilling result of a good union can be the emergence of children to the family. As a stress-reliever, there are many beneficial diversions: helping to educate; enjoying interactive conversations on serious and amusing subjects, and joining in recreational activities. In this author's case, being a father of two boys enabled many chances to train and encourage them in swimming, Little League baseball, sailing, horseback riding, backpacking, playing golf, and traveling to exotic places.

World Series at Giant's Ballpark
(courtesy of Bruce Richardson)

These events were so diverse and different from work frustrations that they provided excellent removal of all work strains. These experiences also expanded the life attitudes of those boys: one became an excellent golfer trying for the pros, and the other carries a lifetime adoration of the San Francisco Giants, his first Little League team. Both have retained steadfast determination, self-reliance, maintaining a good balance between work and pleasure, and respecting the differing cultures throughout the world.

Golfing Challenges

Starting at the age of thirty-eight to learn the mysteries of golf is tough; most people enjoy agile physical skills during youth, but that tapers a bit in midlife. Fortunately, this effort was just in time to incite the sons, and as mentioned above, one son became a lifetime competitive expert. Some rounds were played in exotic global locales: Scotland, South Pacific, Kenya, Morocco, Mexico, Canada, Caribbean, Hawaii, and many sites throughout the continental United States. The noble game also provided many opportunities for bonding interfaces with other Hughes executives, those in other companies, as well as challenging domestic and overseas customers. These friendly join-ups took place in pleasant and invigorating atmospheres. Another advantage was to experience the natural beauty of the landscapes and surroundings of many exotic golf courses. It is a privilege to have checked off 370 different courses: some good, some mediocre, some very difficult, and many with historic and symbolic reputations.

Golfing Vista

Sail Boating

Another exciting active adventure is handling sailing vessels in fair and foul weather. To do this properly, one needs to learn the behavior of sails, booms, yards, tillers, control of heel angle, use of compasses, and the skills of anchoring. There is a plethora of spirited words to learn, including sloop, ketch, yawl, mizzen, larboard, close-haul, beam reach, tack, yaw, jib, spinnaker, jibe, luff, in-irons, come-about, broach, brace, kedge, capstan, and dingy. Fun!

Hobie Cat

Of course, it is essential to know what direction and speed the boat will move with changing wind force and direction. One should also obtain a boating permit from the Coast Guard or state waterways departments. Various agencies operate instruction courses in safety, boat handling, rules of the road, navigation, traffic awareness, and communication. There can be great joy in owning boats of different types—Sabot, Omega, Hobie Cat, and Cal 20—or renting sloops of 32-foot hulls. We enjoyed many weekend family cruises to Catalina and Santa Cruz Islands, as well as adventures on lakes in the High Sierras. You can bet this was extremely diverting from the many difficulties at work.

Wilderness Adventures

Stimulating experiences came from joining each son as they made their advancements and awards in the Boy Scouts. There were many difficult outings: whitewater rafting California's American River and Wyoming's Snake; week-long canoe trips on the Green in Colorado, and the lower Colorado River near Yuma, Arizona; camping and hiking in Bryce, Zion, Yosemite, Joshua Tree, Mojave Desert, Anza-Borrego, and Death Valley. We topped the crests of San Gorgonio, San Jacinto, and many of the other tall California peaks. The traverse down and back up the steep cliff-side trails descending the Grand Canyon meant a vertical mile drop in seven miles. Departing the top in snow, the hikers gave way to mule trains carrying tourists, crossed the raging Colorado on an ancient bridge, then camped overnight in summer heat on the canyon floor. After another night camping partway up another trail, the finish was greeted by a blinding snowstorm. A real manly adventure! The shocker to our scouts was being passed halfway down by a group of Girl Scouts traveling fast because they were going to finish in a single day the down-and-back hike. Ouch! Most challenging but quite enjoyable were many week-long backpack hikes in the Sierra Nevada mountain range.

The adults carried forty pounds and the scouts thirty pounds. These burdens gave sufficient equipment and sustenance to make the trek. The trails were usually easy to follow, but distances to march each day took judgment, as well as being sure a usable campsite was available before

Half Dome, Yosemite National Park

sundown. Many sites, for good environmental protection, cannot be used for tenting and cooking. Nowhere can open fires be used, so we heated our chow on portable propane gas stoves. It was important to plan and bring enough clothing and down sleeping bags to overcome nighttime temperatures below freezing. The jaunts usually began at about 3,000 feet altitude (transportation autos from home can be parked for a week at the start point), ascended over 8,000-foot passes, and continued along ridgeline crests at altitudes as high as 13,000 feet. Each day was a ten- to twelve-mile hike. We conquered Mt. Whitney, the highest in the lower States—14,505 feet as of 2017, ten feet higher than it was in the 1970s. The Half Dome assaults ascended one mile up requiring a sixteen-mile round trip. It was particularly challenging and worrisome to the Scout leaders: its tabletop crest has a rock ledge protruding fifty feet over the sheer front face of the dome. There are no safety cables or rails; imagine trying to corral a bunch of wild boys from inadvertently cascading down 4,800 feet to the rocks below if they leaned over too far!

Seeing the World

A work escape that all can relish is recreational travel. Affordability may be an inhibitor, but much can be garnered even from visiting an adjacent town or some of the magnificent National Parks in the United States. If possible, much mental expansion and insight can be gained from journeys to foreign lands. It is best to try to appreciate its culture by mixing with the locals, learning their history, and digging into what may be very different housing, foods, music, and ceremonies. Many Americans do not do this, losing much joy and often being despised by the local

Ahu Tangariki on Easter Island

residents. Intriguing travels can be done most of one's lifetime, especially after retirement. Luckily, the author visited 114 different countries for both business and pleasure. Some of the unusual business encounters are described in chapter 7.

Everyone who spans the globe has tales to tell and can recall which locales best suited their fancy. To choose favorites is most difficult and greatly influenced by individual taste. Here are some toppers outside the United States, sequenced by most fascinating within each theme. It would have been easier to list twenty favorites in each category rather than observing page limits. Historic wonders: Italy, Iran, Egypt, Japan, Rhine River. Pleasing cultures: Mexico, Philippines, Australia, Italy, Morocco. Visual vistas: Switzerland, Canadian Rockies, Indonesia, U.S. Southwest, French Polynesia. Astonishing wonders: African safari, Iguazu Falls, Machu Picchu, Teotihuacan, Easter Island.

Resort Escape

As mentioned above, vacation times should be sacrosanct to help refresh employee vitality. Since world travel may not be of interest, it is soothing to find an escape location that can be revisited periodically. Many sites may appeal to individual tastes, recreational leanings, comfort of weather, as well as remoteness from the daily pressures and humdrum of the workplace. Options for people in Southern California include Catalina, Pebble Beach, Mammoth Mountain, Palm Springs, Hawaii, and Mexico. Some have history and culture interests, two have skiing and hiking, most have golden beaches, and all have excellent golf and tennis opportunities. Securing a periodic but long-term arrangement can be done by scheduled rentals, contracts with time-share outfits, or temporary house swapping of your home with one elsewhere. For those who can afford it, it is a great pleasure to purchase a condominium or house located in a favorite escape location.

The author discovered, through use of a vacation-timeshare, Sunriver, Oregon. It is an ideal resort area, with a lodge, many condominiums, and individual houses. Nearby are three fine golf courses, tennis courts, swimming pools, bicycle trails, horseback riding, canoeing, and skiing in the winter at nearby Mt. Bachelor. The area is magic: 4,500 feet altitude, high desert terrain sliced through by the lively Deschutes River, and awesome views of the snow-capped Cascades Mountain Range. A mountain cabin with northwest architecture was created among the pines, right on the river, with a spectacular mountain view, and only a mile from the small airport.

Sunriver Escape

Dreaming of and Becoming a Pilot

Being a part of the aerospace profession gave continual exposure to the technologies, history, and participants in all facets of aviation. As the urge arose to become familiar with our military aviator customers, it was most fortunate to experience many dynamic flights in combat fighters at Point Mugu and Miramar Naval Air Stations in California. As described in chapter 6, this was thrilling, but also added further inspiration to begin personal private aircraft piloting. Learning to manage a flying machine is relatively easy after a lot of effort. Being a "pilot" means safe flight, following the rules, and understanding the physical details. 'Tis far more difficult to be an "aviator," wherein the person and the machine blend as one in instincts, physical actions, sensations, and responses.

Piloting your own aircraft is certainly a diversion from work pressures, but it requires much training, practice, diligence, safety awareness, constant vigilance, and lots of stamina: a very different kind of stress. It is exhilarating to be at the controls, but it is never truly relaxing. The fantasy to fly started at age five when our family went down to the Honolulu Harbor to watch the very first commercial crossing by air of the Pacific Ocean. The "Pan American Clipper," named after the handsome speedy tall ships of the nineteenth century, left Alameda, California, and eventually reached Manila after four refueling stops. What a wonder for a small boy to see that huge flying boat descend from the skies and gently touch down on the calm sea. "I've got to do this someday" still rings in my head. Further excitement occurred witnessing groups of Japanese combat aircraft swoop overhead to make the 1941 assault on Pearl Harbor.

During the next three years, our family saw flocks of many types of military aircraft being ferried to combat in the Western Pacific. Such excitement bred further incentive to harness that throttle and stick.

That second abode in Sunriver, Oregon, was eight hundred miles from home. Getting there by driving or commercial flights was quite tedious and time consuming. Why not just leap up there for weekend escapes using a personal flying machine?

Needed was learning how to personally master the flight controls with confidence and safety. An excellent flight instructor at Camarillo Airport allowed a solo performance after six hours of training in the air. Soon thereafter, a pilot license was awarded by an FAA examiner. It then seemed wise to start learning instrument flight in foul weather. Graduating from that tough and grueling training imbued enough confidence to acquire a "hot shot" airplane, the Mooney 206.

Mooney TLS
(personal files)

After enjoying it for five years, the Mooney TLS appealed as a replacement, cruising at 250 miles per hour and booming along at 25,000 feet altitude. That is well above most small aircraft traffic, and below the airliner routes. This wonder felt like a commercial reincarnation of Britain's memorable and handsome *Spitfire* fighter of the 1940s. This new TLS gem delivered pretty fast journeys to Oregon. One return over those eight hundred miles took only 1½ hours due to a very lively tail wind. Other flight ventures were cross country to Florida, several tours of twelve western states, and three jaunts to Vancouver, Canada. One silly objective was to land and takeoff from every civilian airport in California. Checked off were 158 of 350 sites that were currently in service. There were several bouts with terror caused by icing, sudden visual obscurity, severe turbulence, and an engine failure. Lucky to get through all that danger! Those scary events finally culminated in executing a bad landing at Eureka, California. There was an adverse side wind and a very short runway. Touchdown left only half that distance to stop. Unable to come to a halt, the attempt to recover back to the sky was too late. A sand dune crunched that beloved Mooney. This "pilot" almost was done-in by seven major injuries, but fortunately a rescue group raced this survivor six miles to life-saving doctors at a city hospital. After release two months later, it became possible to walk again. Wisdom ruled this as the finale of personal piloting, after exactly one thousand hours recorded in the logbook. Not the best of skill records, but it was great fun and a positive fulfillment of a lifetime dream.

And Then There's Sports

Blissful escapes can arise through active participation in physical sports and competitive games such as contract bridge and chess. These usually do help stimulate brain agility, lively bonding with peers, and the excitement of competition. The environments are also measurably different than the workplace, many times outdoors instead of sitting at a

Trojan's Home
(personal files)

desk or pondering in a laboratory. There are numerous diversions that are done passively, including book reading, movies and TV, stage and musical events, and informative lectures. Many fans also dote on rooting for a professional sports team, attending games, cheering them on TV, and keeping seasonal statistics. A more personal thrill can come from closely supporting some sporting teams as college alumni. Some universities even have an exciting national competitive ranking and a raucous school spirit that lasts a lifetime. Who in this country does not quake when hearing the rally cry "Fight On!" for the inimitable red-and-gold Trojan football team?

Relishing Antique Autos

Another fine stress reliever is to select and enjoy a well-functioning antique auto. The idea is to make short trips around home, pretending to wander in this time machine into a bygone glamorous era. The vehicle could also share a physical piece of history with the neighborhood. A long admiration of the British, their equipment concepts, and the unusual devices they employed during the Battle of Britain steered the final choice. The artistic shape and sound of one gem formed a romantic attachment. The English company Morris Garage had created a whole series of appealing models; the title MG was preceded with its year of manufacture, followed by letters designating the model's series. The real beauty was a *'48 MGTC*. That version was the last configuration of that type before significant modifications were made in the TD model to match the U.S.

Author's Cherished Beauty
(personal files)

market. Many hours of joy came from this selection, and it still is most sad to have parted from it after ten years.

Paying Back

As one gains more experience and wisdom, it can be very therapeutic to plan and perform several ways to pay back the people and agencies that formed your professional foundation. Usually resulting will be a sustained feeling that good outfits will thrive and improve. Self-esteem is nurtured by knowing you are helping future generations to grow as you did, and likely will do even better. How to do this? One is to give your time volunteering to complete physical tasks, mentor others, teach classes, or suggest operating improvements. Another is to donate financial assets for facility improvements, educational scholarships, purchase of supplies, or even sustaining of daily operations. Any of these will earn donators emotional benefits, as well as earn some income tax offset.

San Francisco Maritime National Historical Park
(photograph by Chris J Wood)

Being selected to mold policies and render intellectual support can be a great honor. It gives the chance to directly apply wisdom to influence the future course to be steered by nonprofit groups. Such an opportunity can arise from membership on a formal board of trustees. Positive enthusiasm in participation by newcomers is welcome, but there will be many frustrations to the neophyte. Charitable organizations depend on donations, so continuing romance and charm must be applied in all relations with any financial sustainer. Some donors may also be board members, may have been one of the outfit's founders, or perhaps could be a current leader of the local community. Sometimes, when a bright new idea is presented, these well-rooted old salts place significant barriers to implementation. Similar impediments may also be encountered in a business corporation, but usually the opposition can be swept aside because competitors will likely leap ahead if the firm is immobile. Not so in nonprofits since these folks cannot be easily cajoled to remove their roadblock, and other board members are fearful of offending them. There also can be lethargy because many of the participants are volunteers, not driven by desires for their own professional progress. Once a new board member gets used to this molasses operating style, ways can be found to maneuver and persuade the recalcitrant to not impede progress expected from worthwhile actions. A business executive, normally proceeding chop-chop-chop, can learn political finesse and patience, both valuable skills.

Good work also may earn a meaningful formal award from the charitable organization. These are rewards gaining public acclaim. It was an honor and privilege to serve on six boards: Tufts University in Medford, Massachusetts; the Professional Golf Association in New Jersey; the Santa Barbara Zoological Gardens; the High Desert Museum in Bend, Oregon; the Estrella Warbirds Museum in Paso Robles, California; and the San Francisco Maritime National Park Association.

11

SWORDS TO PLOWSHARES

WHEN MANY CREATIVE organizations were in place in this nation, companies searched for new commercial product lines. Spurring continual inventive progress requires long-term funding to sustain sufficient talent and working facilities. Internal asset sources can and should be planned as a high executive priority. But, as mentioned in chapter 8, large government investment frequently provided the rich technical breakthroughs that formed the foundation for companies to create many life-improving devices now commonly used worldwide. Most advances in electronics product lines during the twentieth century's last half were funded by the United States and other nations. Public support was enthusiastic, because of the urgent need to assure survival from the economic and political threats posed by the Soviet Union. Project objectives were military superiority; comprehensive and instantaneous intelligence gathering; secure and unlimited communication; as well as scientific endeavors. Other markets existed, however, that could yield corporate growth in revenue and earnings. Internal research and development sources were then used to exploit commercial products.

In the following beneficial product lines, those creations by Hughes were the result of actively applying the operating paradise described in chapters 3 and 4. Many of these tools and toys are today taken for granted by the public. Their conception and how they were brought to fruition should be of great interest to everyone, most especially to leaders and political activists. They should and must find and energize the best methods to continue such technology progress into future generations. From a national standpoint, this is essential for our survival and prominence in the completely interlinked but competitive global economy.

Instant Thinking and Always Remembering

When we see the word **computer** today, what usually comes to mind is the personal computer residing on many desks, or the laptops carried by many travelers. However, almost all of the computers that serve us are not so visible. They are within cell phones, toys, iPhones and BlackBerry devices, wristwatches, and home appliances. An automobile may incorporate dozens of computers, an aircraft hundreds, and a military or space system thousands.

These machines manipulate data according to a set of instructions. Mechanical devices that do this date back to the Antikythera mechanism, an astronomical calculating device made by that Greek philosopher in about 100 BCE. More recent were the nineteenth-century Jacquard loom and the World War II Norden bombsight.

Today's computers evolved as an amalgam of successful ventures by many individuals and teams. The four founding blocks were conversion of analog information to binary digital numbers, semiconductor devices, microminiaturization, and software programming languages. The rapid

The Wonders of *ENIAC*
(courtesy of Columbia University Computing History Center)

modern development of computers began in the 1940s, efforts greatly stimulated by survival urgencies during the global war. Early systems whose behavior could be altered by inserting instruction changes include the 1941 German electromechanical Zuse Z3 and the 1944 British all-electronic Colossus Mark 1. However, the first fully programmable electronic computer, and the best publicized, was America's *ENIAC* (electronic numerical integrator and computer), designed at the University of Pennsylvania under a 1943 secret contract with the USA. After its public revelation in 1946, news media colorfully proclaimed it the "Giant Brain."

Fashioned with 17,468 vacuum tubes and many thousands of individual resistors, capacitors, inductors, and transformers, it occupied two large rooms full of tall equipment racks. Performing 5,000 additions or 385 multiplications per second, it had no separate memory and could operate only twenty 10-digit decimal numbers at once within its circuitry. Later generations of computers rapidly improved speed, memory capacity, and amounts of differing functions performed simultaneously. Advances included great reductions in computational errors, less power consumed and heat generated, as well as improved reliability and life. Sizes shrank an astonishing degree. Large vacuum tubes were replaced by miniatures, then by solid-state transistors, and eventually by tiny wafers containing thousands of interconnected components, known as "integrated circuits." Memory types progressed from numerous vacuum-tube storage rings through paper card aggregations, spinning magnetic drums, and finally to magnetic cells infused on thin wafers, as well as laser discs with embedded digit sequences. What had taken two huge laboratory rooms in 1946 became the size of a half-slice of bread by 1995, with several million times more memory and operating speeds jumping from hundreds to billions of computations per second.

Computers now usually include a high-capacity storage memory, an active temporary working memory, an information processor, and a set of pre-planned operating instructions called a "program." In addition, there are input-output connections that bring instructions and data into the computer and that make the results of its calculations available for display or use elsewhere.

These machines can accept and process a great variety of information, including words, images, tables of data, streams of signals from sensors, and countless equations. All of these are translated into sets of binary digit numbers, which can then be mathematically processed according to the instructions in the computer program. For example, a photographic image may be encoded as a set of numbers that define the brightness and color of each of the millions of pixels (picture elements) that make up the image. This is then sent to a user's machine containing a search program with memories of known objects. It then analyzes the scene and highlights only the features of interest, such as terrain contours, buildings, land, or vehicles. Future planning for the area surveyed can be done more easily: likely sites for oil exploration, flood control, effective routing for new roadways, or changes for environmental improvement.

GPS Can Find You

Less than twenty-five years after worldwide satellite communication became possible, another remarkable composite of electromagnetic devices established the phenomenal worldwide *Global Positioning System* (GPS). This enables anyone equipped with a signal receiver to know precisely their own current location and the earth position of other objects, and it enables calculation of the best route to reach desired destinations. Begun as a U.S. defense project in the 1970s, this amazing benefit to civilization became fully operational by 1985. The ultimate configuration is attributed to Bradford Parkinson, Roger Easton, and Ivan Getting. A network of twenty-four to thirty-two satellites orbit 12,600 miles above the Earth, each circling twice daily. At least four space birds are placed in each of six separate orbital planes equally spaced around Earth's axis. All these rings are inclined 55 degrees from the equator. In each plane, an orbiting satellite is followed by another within 1 to 3⅕ hours. This random spacing always assures that at least six are visible from any spot on Earth. All satellites have a precise time clock updated by a ground master station. Clients have a passive device no bigger than a matchbox that observes signals from four satellites; one assures time correlation. At an identical sample time, the three others each transmit a distinctive pulse. The receiver computes the distance to each by measuring the time the signal needs to reach the receiver from

GPS Satellite Galaxy

that bird (as mentioned earlier, radio waves travel at the speed of light: 186,000 miles per second, and corrected in GPS for the effects of satellite velocity). The client's location is at the only spot on the Earth's surface where those measured distances from the three sampled orbiters meet.

Seeing Time by the Numbers

During its early expansion years, Hughes made many attempts to enter the highly competitive commercial world, targeting both consumer and industrial customers. We made substantial inroads into the industrial sector but were quite unsuccessful in the consumer sector. Selling directly to millions of individuals can be very lucrative if large quantities can be rapidly produced at very low cost. Hughes had no experience in this kind of manufacturing, nor in forming the marketing network needed to survive in a world of fickle consumers with quickly changing desires. As soon as we built a newly invented device and placed it on the consumer market, another company quickly copied it (outside the reach of our patent protection) at a much lower cost. To grab maximum market share, that competitor set its price well below its unit production cost; its well-established marketing network drove our product out. That competitor then increased the price as it became the dominant or single source, and quickly repaid its earlier investment loss. Even if we signed up with a major distributor with consumer marketing prowess, our top executives were unwilling to take on the unknown risk of establishing the needed production facilities. They decided we were better served by focusing our venture funds on research and development for government systems. Nonetheless, the company did invent some pretty ingenious products across a wide variety of technical disciplines, as seen in this chapter. Perhaps if there had been a string of early successes, confidence and maturity would have resulted in quicker investments in, and entry into, consumer ventures. Speed is of the essence in that bubbling marketplace.

A good example was the computer-driven *digital watch* with its innovative liquid crystal display. These timepieces were not like anything that had existed before, and the public's enthusiasm blossomed into fruitful sales and earnings. But the company was slow to set up a low-cost, higher-output manufacturing plant. Quickly, large-scale domestic and overseas outfits made close copies and willingly invested in automated production lines; their wide distribution nets soon captured the frenetic market. Patent holdings were insufficient to stem the tide, so we abandoned this marvelous small product in less than two years, making $50 million and then losing $50 million through pricing pressures. (Some of our competitors also retreated much later after losing their shirts.) It had become apparent that we could not sustain ourselves in this particular consumer market. We had the brainpower competency to jump ahead with new designs, but could not compete in what the buyers perceived as a jewelry business, where, not surprisingly, fancy appearance counts more than the product's operating features.

First Digital Watch
(courtesy of Jeff Grant)

Operating in the Dark

An ancient frustration to humans was finally remedied with night vision using **infrared** (IR) sensing. The term "infrared" is used to define a segment of electromagnetic radiation in the form of what is usually perceived as heat. The word suggests that the thermal energy frequency is just below (infra) that of the visible color red. Sometimes objects show both IR and visible light at the same time, such as a candle flame. The thermal frequency band is the next parcel above that of many varieties of radar emanations. Infrared radiation is invisible to the unaided human eye. By building sensors that extend our vision into the infrared, we can often observe through haze, light fog, smoke, and dust. Because object emanations at IR frequencies differ from those at visible frequencies, with new sensors we may be able to perceive objects hiding in shadows or concealed with camouflage. Better yet, all warm objects, including people and vehicle engines, emit infrared radiation, so they can be easily seen at night using equipment that convert IR to visual displays. Fortunately, such systems are passive: they receive meaningful signals, but don't emit energy that reveals their location to those being observed. Hughes became the world leader in developing and manufacturing military combat devices exploiting these phenomena. Products included targeting for riflemen; artillery, tanks, and fighter aircraft; missile guidance; as well as hostile ICBM location and destruction. Many of these included single-spot detection, tracking, and high-resolution imaging.

Early IR systems in late World War II, as well as in *Sidewinder* and *Falcon* missile seekers in the 1960s, did not provide complete images. They used a single or just a few detectors pointed in the general direction of a hot target such as an aircraft or a tank, and then were moved at different angles until it pointed directly at that target. Much ingenuity was sought to greatly improve performance. Needed were better sensitivity, superior control of physical dynamic positioning, scanning accuracy, and the ability to save and process data to form useful images.

Our company made practical many unusual sensor material compounds that were far more sensitive than the lead-sulfide materials used since 1946. That relatively simple material had an advantage of operating at ambient temperatures. The new detectors required artificial cooling to achieve their best responses to incoming energy. We created many unique low-power demand coolers using liquid gasses: argon for -309 degrees F, nitrogen for -320 degrees F, and helium for an incredible -453 degrees F. Long-term system functioning was limited by the storage space for these liquified gases, since they could be quickly depleted. To solve this, especially for spacecraft, several varieties of mechanical refrigerators and cryogenic engines were developed.

Numerous successes in discovering new compounds to fabricate detector wafers finally culminated in a mix of mercury, cadmium, and telluride (HgCdTe), usually called "mercadtel." Finally perfected in the early 1980s, it had been a DoD project for many years of frustrating experimental disappointments, prompting an official to remark, "HgCdTe is the detector material of the future and it always will be!"

Upon its becoming practical, it was soon found that detectors using these compounds could be tailor-made to respond anywhere within the entire IR band by varying the amounts of cadmium and telluride. Its rapid response to incoming energy enabled sophisticated scanning images of large areas. Its -320 degrees F cooling demand was easily done with flasks of liquid nitrogen.

Non-military usage of this very advanced compound was primarily for capturing observations of the universe to new distances and differing perspectives than those using visible light. Our first commercial venture, using lesser technologies, was a tool to examine buildings and determine

Thermovision Image of House

weak areas where existing insulation was ineffective. Such IR images were converted to multi-colored displays so repair personnel could easily see where repairs should be made. Other devices were for police agencies to enable nighttime searches for miscreants, and for firefighters to zero-in on hotspots needing priority quenching. These tools could be used afoot or in cars but were more useful when mounted in helicopters. These images had sufficient sensitivity to be effective in complete darkness, and enough object resolution to distinguish body shapes as small as a cat.

Observing Earth's Wonders

Landsat is a satellite that continually creates images of the terrain beneath its orbit track. The first was launched by RCA in 1972, and the eighth began operations in 2013. Each bird had a fully functioning lifespan ranging from three to thirty years, thus providing the longest continuous source for this form of scientific data. The principal sensor systems are called Multispectral Scanner and Thematic Mapper, developed by the Hughes Santa Barbara Research Center. Pictures 115 miles square with 40-foot resolution can be captured by ground stations throughout the globe. Sequential images, taken sixteen days apart, can be compared with stored ones to measure environmental changes over protracted periods. Each scan records separate spectral data, from infrared through visible colors to ultraviolet, giving multiple data forms to analyze the actual conditions of the landscape.

The information has provided enormous scientific information to understand forestry harvesting, wildfire vulnerability, global warming, the condition of fisheries, accurate three-dimensional terrain mapping, and glacier shrinkage. Very meaningful to the public are weather status and forecasting, pollution conditions, agricultural improvements, and natural resource management.

The Bahamas from *Landsat*

Making Light Do 'Most Everything

Lasers, found in many diverse applications, came to practicality in 1960. In the next fifty years, there were 55,000 U.S. patents issued for uses in every field of human endeavor. Talk about utility and adaptability! The word "laser," short for "light amplification by stimulated emission of radiation," was first commercially introduced in 1974 as a barcode reader at supermarket checkout. Text readers appeared in 1978, compact audio disc in 1982, printers in 1983. Then came directional pointing flashlights, three-dimensional holographic pictures, DVD movie discs, and dramatic citywide light shows on buildings. Secure high-speed long-range communication links emerged using glass fiber optics transmitting laser light. In manufacturing, precise cutting, bending, marking, burnishing, and welding could be done faster and with high quality. Medicine now greatly benefits in microsurgery, diagnostic procedures, and many forms of physical therapy. Scientists use lasers for comparative wavelength measurements, precise positioning, and cutting and burnishing chemical films.

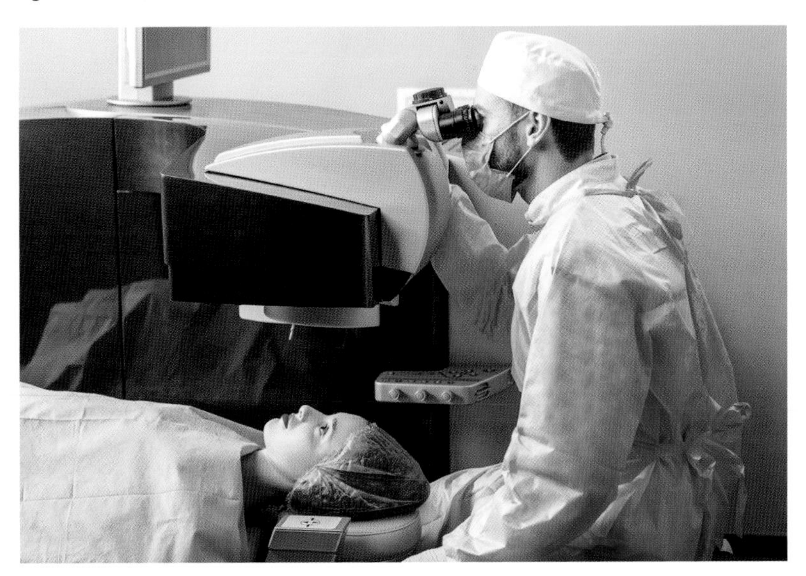

Laser Eye Surgery

Dr. Albert Einstein in 1917 described the concept of stimulated emission as a physics phenomenon. For the next thirty years, laboratory experiments showed its feasibility, and in the 1950s Dr. Alfred Kastler demonstrated a method called "optical pumping" to increase the output of radiation. Charles Townes of Columbia University in 1954 showed such emissions in the millimeter-wave band (a device called a "maser"). For this, Townes was awarded the 1964 Nobel Prize in Physics. He also provided a good step forward to achieve the future laser successes. Dr. Theodore Maiman at Hughes finally achieved in 1960 the first operating laser in the visible light waveband. The device used a pure solid crystalline ruby whose atoms were known to emit visible light when stimulated. Its inch length was shaped into a 0.4-inch cylinder. A nearby light flash got it started, and a surrounding electric coil provided continuous excitement. Mirrors at each end bounced the light back and forth to build its energy level. One mirror had a tiny hole at its center to emit a portion of the amplified light to emerge for use. The ¼-inch-wide pure color beam would expand to only 1 inch in a mile, or to only a 1-mile circle if the beam traveled 240,000 miles to the Moon, as measured by the Hughes *Surveyor 7* in 1967. Many other materials were found later that emit light of every distinctive radiation type down to lower frequencies of infrared and microwave, as well as upward to ultraviolet and X-ray.

Many applications project a continuous uninterrupted beam. Others needed short pulses of high peak power to measure distance, as is done by radars. The laser beam must be rapidly switched on and off. R. W. Hellwarth and F. J. McClung at Hughes perfected Q-switching by inserting a thin-film of nitrobenzene in front of the mirror at one end of the laser cylinder. This insert uses the "Kerr effect," discovered in 1875 by Scotsman John Kerr. When activated, the cell stops normal laser operation by rotating the polarity of the reflected beam and quenches further amplification. Switching times can be extremely fast and easily controlled for power management and exact timing intervals to ensure precise range measurements. The first prototype achieved pulses as short as an eighth of a microsecond and a peak power of 600,000 watts, startling performance from such a tiny device. Today, lasers vary in power output from DVD's one hundredth of a watt to the staggering 1998 emission at the Lawrence Livermore Laboratories at a peak power well over 1 trillion watts!

Hughes did not participate much in commercial applications using lasers, but became the global leader in manufacturing military laser devices. Entirely new combat methods sprang forth with lasers: high resolution surveillance imaging, target designation, accurate range-finding, target illumination for guided missile homing, aircraft pilot heads-up displays, and projection of destructive high-energy beams.

A scary surprise occurred to local Culver City residents when they witnessed what appeared to be the dreaded arrival of aliens using UFOs. We were testing a gimbal scanner for a laser tracking device. The plan was to measure the deviation of the tracker as it directed the laser beam to trace a mile-long straight line painted on a wooden fence across the company airstrip. The software went haywire. The beam rapidly darted randomly from one cloud to another. The locals were terrified, convinced these were UFOs, and called police and fire stations to solve the impending problems. We could not respond to any queries since this was a highly classified military project. The first range-finding device was perfected by Dr. George Smith using Q-switching aided by a rotating mirror. It boasted an incredible measurement accuracy of ¼-inch in ten miles. One of our most successful large-scale manufacturing programs was a motion-stabilized IR and optical sight tank fire control system *(TFCS)* with an integrated laser range finder and target illuminator (see chapter 8). In 2017, airborne laser systems that terrain-map with range measurement were in use to determine snowpack depth in the Sierra Nevada, monitoring California's adaptation to prolonged drought conditions.

Information Flow Everywhere

Worldwide **communication** has been made possible by satellite data relay systems. Hughes branched out to conquer this challenge as a means of saving its radar design staff. This business group's very successful developing and manufacturing of airborne radars for the military appeared to be declining. Federal budgets were prioritized toward the Vietnam conflict, and heavy future budget cuts were forecast. The idea was to use radar frequency data links aimed upward at a "fixed" spot in the heavens and have sites in remote locations catch the data as it was relayed back to Earth.

The inspiration of a communication satellite was first described in writing before the technology was adequate to accomplish the difficult task. In 1928, Slovenian rocket engineer Herman Potocnik, discussed the idea in *The Problem of Space Travel: The Rocket Engine.* Shortly thereafter, Britain's Sir Arthur C. Clarke further ignited public fantasy with his many science fiction stories. In 1945, after serving World War II as a radar expert for the RAF, he defined how a workable satellite communication system could be implemented.

To attain geostationary orbit a spacecraft must be boosted over the Earth's equator to 22,236 miles above the surface and remain at a speed of 6,489 miles per hour. This velocity sets a centripetal outward acceleration force that exactly matches the Earth's gravitational pull, just like a child's whirling a rock attached to a string. The satellite's movement traces exactly over an equatorial spot directly beneath itself. That location allows ground-station data streams to be relayed to many receivers placed anywhere within one-third of the world's surface (the portion that is visible from the orbiter). Antennas both at the source and client could be fixed in their look-angle rather than having to track a satellite with a lower altitude orbit. Incidentally, AT&T had warned that if such relays were used for telephone service, clients would be frustrated by the short time delay for the signals to go up and down. This splash of cold water was intended to forestall competition with its profitable undersea cable business. That effort failed, since users quickly adapted to the delay during their overseas conversations. Equatorial placement does limit visibility from regions beyond the Arctic and Antarctic Circles since the spacecraft appears below the Earth's horizon.

When several satellites are placed in the same orbit circle, and relay data at the usual high frequencies, they must be spaced at least 3 degrees apart (a physical distance of about 930 miles) to avoid overlapping antenna patterns and to ensure proper linkup with the correct ground station. When using much longer wavelength like UHF, only one satellite per ocean basin was permissible to prevent mutual interference.

The first geostationary bird, called *Syncom*, achieved operating orbit in 1963. The highly motivated and inventive Hughes team was led by Don Williams, Harold Rosen, and Tom Hudspeth. The creation was a 28-inch cylinder, 15 inches high, weighing 86 pounds. Its exterior provided enough area for the number of solar cells needed to satisfy the continuous power demand. Flight attitude was stabilized by spinning the cylinder at 160 revolutions per minute to act like a flywheel. A very simple device was used to keep the spin axis parallel to the Earth's. Two small clamshell canisters, each containing a solar cell, were placed on the spinning exterior. A small slit permitted the sensors to glimpse the Sun for a brief instant and thereby provide an electrical pulse. The base-reference sensor had its slit parallel to the satellite axis and peered 90 degrees outward; the other had its slit at an angle to match the Earth's normal tilt relative to the Sun. Pulses from both were sent to the ground for computing corrections that would assure that the pulse from the tilted sensor recur at the same instant as the one from the base-reference sensor; otherwise, the satellite axis was not aligned correctly. Corrective commands sent on the uplink then triggered *Syncom*'s small alignment thrusters to

HS-367 Communication Satellite

fire until the two pulses came at the same moment. Ground computer calculations always adjusted for the known wobble of Earth's axis.

Following this great initial success, a booming military and commercial demand for sophisticated ground stations, satellite-mounted relay modules, and ground receivers provided a strong business boost for the Hughes radar sector, fulfilling the original strategic goal. Even more rewarding, it became the beginning of a leap to global leadership in satellites and great successes in spacecraft scientific exploration. Earnings from other corporations using our patents for spin-stabilization, Sun sensor attitude control, and minute radar-frequency relay modules for spacecraft were lucrative. For many years, the company had 60 percent of the world's operating satellites in orbit. Most sought after by customers was the *HS-367* Communications Satellite.

Coaxing Consumers to Connect by Space

Stable and lasting space-business return could be captured by selling services directly to the civilian populations in large countries. To prepare for this, Hughes grew its capability to design and produce all the space and ground equipment, staff the operating stations, and market to millions of

HS-601 DirecTV Satellite
(courtesy of UNLV)

private consumers. What emerged was *DirecTV*. After we invested about $200 million, customer service began in 1994. The bird relayed over three hundred independent television channels to individual homes and buildings, as well as 48,000 telephone connections. Subscribers grew rapidly and within ten years reached over twenty million. In 2016, the venture was purchased by AT&T for $ 48.5 billion. That was a remarkable growth in twenty-five years of a relatively small investment. (Chapter 9 describes the business and financial strategy to establish and succeed with *DirecTV*).

Initially, to spur signups and thus proclaim a market surge, we provided the adapters free to subscribers to assure compatibility with their existing TV sets. The first ground station, in Colorado, provided the digital uplink at a rate of one billion digital bits per second. Our public offering quickly captured the U.S. market, and within a short time reached fourteen million subscribers, more than were served by any surface cable company. Many improved versions of HS-601 birds have been placed in orbit in the last twenty-five years. Each now weighs over 20,000 pounds. Seven by 50-foot solar panel arrays provide enough power for sixty transponders to relay 10,000 watts of digital data streams.

Billions of Ringers

Cell phones are now in continuous use throughout the world, with devoted users numbering in the billions. Also, in parallel, the Internet with its even larger information rates and extraordinarily large user base has evolved to a point where there is now a mutual dependence of Internet functions like email, which often utilizes the phones cellular system. It is most difficult to even visualize the complexity of implementing a network architecture, embodying the enormous bandwidths and protocols that connect all the participants, with the assurance of continuity, and avoidance of interference for the high rate streams of audio and video data. Critical to do all this are conversion of data streams to digital, compatible formatting rules in the net, signal compression in time, line-of-sight linkage to network antennas, and signal processing devices which operate at extremely high speed. The foundation of the architectural structures now used is a derivative of the work done at Hughes for fighter/interceptor radars, together with the frequency-hopping invention of the actress Hedy Lamarr, described in chapter 5. We perfected the precision high-speed digital conversion of analog signals to digital and created a processing system governed by software. This software could be actuated with its internal memory to enable the hardware system to restructure instantly for the unique modal needs in dynamic situations, such as approaching the target from its tail rather than its nose, examining the object in more detail to obtain identification, and for obtaining high-resolution ground imagery.

Signal processing is the function of analyzing the minute details in the information captured by a sensor. High performers can minutely examine the data content down to a fraction of the distance between incoming wave crests (much less than ¼ inch for 10-GHz radars, and 1 inch for the 3-GHz frequency of cell phones). Imagine accurately sampling these short segments as they rip past at 186,000 miles per second! These fast streams of measurements were then used to pinpoint identification and content. The need in advanced airborne interceptor radars was to sort targets from background noise, determine their course and speed, and even create a structural image to find its most vulnerable spot. As stealth radars evolved, this detailed information permitted changes in the outgoing beams to be modified so as to become virtually invisible to the hostile. Such detail also now perfectly matched the needs of high-speed cell phone mobile instruments.

Early signal processors were analog devices that sampled and typically performed a frequency analysis of the signal stream from the radar receiver. Progress in digital computers in the 1960s stimulated the idea of performing processing in a similar mathematical way. The first challenge was to rapidly and accurately convert the analog signals into binary digits. Conversion was done by measuring the magnitude, phase, and frequency of continuous samples of the electromagnetic wave as it passed a time-clocked position. Measured values were transformed into digital

Sophisticated Cell-network Device

numbers, passed to the processor, and combined into a data stream for subsequent analysis. A number of unknowns had to be addressed and resolved in those early days. Among them were achieving the required sampling rate and conversion accuracy, i.e. the quantization level allowing the most minute of target signals imbedded in noise to be discerned without clipping the peak signal values. Secondly, determining how many samples to accrue for each signal so as to not over-tax the storage and computational load for a real-time on-board processor meeting the frequency analysis needs using the currently available semi-conductor device technology.

Government-funded projects to implement digital signal processing for radars began in 1966. Milt Radant led the innovative Hughes team that created laboratory devices confirming the high-speed data conversion and signal computational requirements. Three prototype systems soon successfully generated real-time high-resolution radar ground imaging streams continuously formed from surveillance aircraft. At ten-miles range, these early maps had resolution down to ten-foot cells or pixels. Demonstration of the other important airborne radar mission of air-to-air search and track quickly followed.

Signal processing requires only very simple repetitive computations, but it has to keep up with the dynamic real-time data input, speeds of 2½ million operations per second in the case of radars of the late 1960s. To do this, one of our systems at that early computer history era galloped along at 225 million operations per second. Today's general-purpose computers usually operate at about 1 billion operations per second!

In the 1960s, the first generation of digital processors had limited performance and lacked agility to change operating functions as each operating mode was hard-wired. By 1971, advances in semiconductor technology led to new generations of integrated circuit devices affording instant switching from one performance mode to another. Complex high-speed digital electronics were managed by software programs. Unique instruction codes for signal analysis and usage could mimic in many ways the volatile leaps forward in software development in the now ubiquitous data processing industry.

Hughes invested large amounts of company funds to derive what became known as the "programmable signal processor," or PSP. Team leaders Dave Lynch and Lee Tower patented the successful design architecture. In 1974, a PSP prototype confirmed the instantaneous reprogramming between multiple operating modes. The first deployed system was in the radar equipping the USN's F/A-18 *Hornet*. We won that competition in 1978, as described in chapter 8.

This very sophisticated design architecture became the standard for all high-performance radars and was soon mimicked to satisfy the complex signal processing demands of mobile cell phone receivers. A brief summary of the architecture is this: rapidly convert radio-frequency signal streams to digital numbers by sampling ¼ wavelengths with precision in time and magnitude; sort those into manageable segments; perform high speed computations using complex equations; identify and separate portions addressed to this particular receiver; convert information back to analog for the user. All this is controlled by an extremely high-speed micro-computer with re-programmable software.

Tiny Is Best

Microelectronic circuits are the essential building blocks of most electronic equipment of today. The term implies pieces of hardware that are very tiny but have enormous operating capability. The smaller, the better; there is a perpetual drive to shrink functions without limit. The original digital computer described previously, *ENIAC*, sported 17,468 vacuum tubes housed in tall racks occupying two large laboratory rooms. It only achieved a puny 5,000 additions, or ⅓ of a multiplication per second. After John Bardeen, Walter Brattain, and William Shockley invented the transistor in 1947, electronics could be fabricated using solid-state wafers with semiconductor elements. Several flat film-like substrates containing differing electronic components submerged or placed on their surface are stacked and bonded into interconnecting layers, forming an integrated circuit. These are usually called "chips," and are further electrically connected with others to form a fully functioning electronic device.

Dramatic shrinkage is illustrated by today's special-purpose machines: 1 billion computations per second can be done with something the size of a small matchbox. Most remarkable are flash drives less than the size of your little finger that store as much as a trillion bits of software memory.

Hughes inventions contributed a number of significant processing methods that aided the continual shrinkage. These are still used today. One is a fabrication method to make thin-film elements interconnect when stacked as layers, forming tiny chips. Another was a submicron lithography for shaping minute circuit sub-layers, permitting large paper layouts to be reduced to less than 0.04 thousandth of an inch! Hughes perfected ion-beam implantation to create unique microelectronic circuit elements by doping semiconductor alloys; the changed material characteristics yielded higher performance, reduction in size, and lower cost of the many elements embedded. Another improvement was to design layers with self-aligned gates, permitting many more transistors on each layer. A new compound called "complementary metal-oxide-semiconductor" (CMOS) became ideal for forming transistors in low power circuits. Resistance to nuclear radiation can be done using silicon-sapphire semiconductor circuits. And a new alloy, gallium arsenide (GaAs), from Hughes permitted fabrication of high-speed, low-noise transistors for radio frequency power amplification, efficient solar cells, and charge-coupled devices.

Powerful Microchips

And Many More

There are many other lifestyle benefits that were derivatives of Hughes technology breakthroughs. Some examples are clutter-free radars that pick out objects in spite of echoes from terrain; radars with electronic scan that avoid physical sweeps of the antenna; complex computer networks sufficient for high performance air traffic control for many nations; high density encrypted radio communications; and precision long-range sonars for shipboard navigation safety. This book's appendix lists many of the unique and high-performance devices created by Hughes Aircraft Company. Details can be found in *Hughes After Howard*. All of the items in this chapter are fine examples of how human lifestyle improvements can spring forth from government development investments that were driven by the crucial need for high performance. Most technical breakthroughs were soon adopted by industries to focus their own talents and funds to create astounding new commercial applications.

12

MOST CHALLENGING CAREER ADVENTURES

As MENTIONED IN the preface, many of the thoughts in this book resulted from active personal participation in the golden age of aerospace. During this time, the author was very fortunate to progress through the seven layers from radar design engineer to corporate president of the leading military electronics organization in the world. Some readers might be interested in hearing about the most demanding and exciting assignments experienced in climbing that ladder. These graphically demonstrate how rewarding and enjoyable can be a lifetime spent in a technology profession. Below are four career milestones.

Moving up the management chain of a large corporation can be very rewarding, with increased compensation and greater prestige, if those are well earned. There will be an expanded understanding of company objectives and operational difficulties, with the authority to simplify the procedures used in your organizational unit. It is likely there'll be an opportunity to increase personal pride of accomplishment. Many individuals do not seek such increases in responsibility because it will carry more intensive work, higher stress, and a possibility of professional failure. Promotions always incur a bit of risk, so accepting them requires confidence, understanding of the work content, and most of all, an inner skill to successfully communicate with and inspire the staff. It is essential that a boss can converse comfortably with those reporting, regardless of their background, job assignment, or expertise.

My upward steps seem to have been on a five-year positive cycle. That was fortunate since in each assignment there was enough time to learn all the peculiarities and tasks of that function, correct the errors I had made, and build a solid foundation of experience to move further up the line.

I do not recall ever dreaming to shoot to the top. My foremost motivation was to perform well in my current assignments and implement improvements in that organization's operating methods. Promotions usually came as a great surprise, giving me the feeling that I was luckily in the right spot at the right time, as that old saying goes.

First Black Triumph

In 1961, the *YF-12* interceptor version of the SR-71 *Blackbird* was slated for air defense of the North American continent. The threat visualized was an over-ocean attack by Soviet bombers bearing nuclear weapons. With these 1,800 miles per hour fighters bearing long-range, all-weather guided missiles, the incoming hostiles could be destroyed well at sea, so that possible ignition of the bad-guy atomic bombs would not occur over friendly landscapes. The Hughes equipment already demonstrated as part of the canceled F-108 *Rapier* Program, the ASG-18 fire control and matching AIM-47 *Super Falcon* missile were ideal for this mission. Its radar could spot small targets well over one hundred miles away, and the all-weather missile could be launched up to eighty miles from a hostile.

This new work was our first "black program," the moniker for extremely high-level government security classified projects, performed in a special manner. Those programs were intended to deny outsider knowledge by hostiles, competitive companies, as well as other U.S. military services contending for DoD budgets. The program's existence, purpose, location, budget source, cost, schedule, usage, and customer objective were closely guarded. All documents were carefully stored; many had to be shredded as soon as possible. Most work areas were sealed within double walls and protected from sound and electronic eavesdropping. A benign form of project identity was needed if many noncleared workers were needed to perform nonclassified portions of the job. Nondescript code names were selected. Some colorful ones were: AQUITONE, OXCART, CORONA, HAVE DONUT, KEYLOCK, TACIT BLUE, AIMVAL/ACEVAL, and HAVE GLIB. Very often, smaller projects were simply coded with number sets, such as 2869, 8936, or 3206. Participants could be allowed to publicly reveal such numbers on team baseball caps for a noon-time softball match. No outsider could decipher those since there was no logical correlation to the project.

People recruited as "insiders" were subject to very detailed and tedious security clearance by the government, sometimes applying lie detectors. Once cleared, that individual had to sign a legally binding lifetime pledge never to discuss any part of the project to anyone without the magic “need-to-know.” This embargo meant that they could not tell their families what they were doing, where they were going on business, or with whom they were interacting. The lifetime limit was only lifted upon formal government notification, which might occur after a twenty-year span.

AIM-47 *Super Falcon* Missile with *YF-12*
(courtesy of the National Museum of the USAF)

Learning how to function properly for many months or years on a black program was difficult. *YF-12* was a lifesaver for two product lines essential to our company. One afternoon in 1961, I became the fourth Hughes person cleared to work this clandestine venture. The others were senior executives Dr. Allen Puckett, Claire Carlson, and Walt Maguire. Black was new for all of us since this was the first program of that type assigned to our firm. I was given **overall engineering authority** to define physical configuration of the entire system’s redesign from its F-108 version and worked with Lockheed to mold the installation provisions. The evening of the day I had been cleared, I was secretly transported to the Skunk Works in Burbank, about twenty miles away, to see drawings of the compartments assigned to our equipment and missiles. Returning to our Culver City home base, I spent until two o’clock in the morning at a drafting board. Using a new secure telephone, I then asked Lockheed to extend the aircraft’s fuselage by twelve inches (the bird design already was one hundred feet in length) to make enough room. By ten o’clock the same morning, Kelly Johnson called back and concurred. This was typical of the fast-track activity that contributed to so many Skunk Works triumphs. Another installation challenge was to internally house four 12-foot-long, 818-pound *Super Falcons* on their ejector-powered launchers. The other strain was to accommodate and environmentally protect about thirty electronic boxes making up the high-performance radar, computer, cockpit displays, and weapon control functions.

About sixty Hughes staff members were security-cleared and worked in a location called "behind the Green Door," a secured area on the second floor of an aging Hughes wooden building. Individuals properly identified by an armed security guard entered that magic gateway and faced another door and a second guard. Since more than one thousand other Hughes people were needed to perform all project tasks, we renamed the ASG-18 the "Standard System," reputedly searching for an unknown aircraft home. Renaming the *Super Falcon* was not necessary since it could be the same size and shape for application in many types of future fighters. My Green Door team mandated ASG-18 design requirements and equipment shapes to non-security-cleared engineers without discussion or debate. Often this nonnegotiable demand could become awkward. Detail designers thought, "If the airplane space is not yet known, why squeeze us so hard?" There was once a major dispute when we directed that the transmitter unit's height not exceed seventeen inches. My position was backed directly by Pat Hyland, the company CEO. No size refusals from any designers occurred after that. This was a meaningful and permanent demonstration of fully delegated authority.

When we completed a full-scale wooden mockup, it needed to be moved to the Skunk Works for a trial-fit in the aircraft prototype. To do this clandestinely, I rented a large truck, marked it with the name of a bogus transportation firm, donned a truck-driver's outfit, and drove the empty vehicle to a Hughes freight delivery gate. The posted security guards, not informed about this black program, had been instructed to let this particular truck pass without question. A crane was used to grasp the large canvas-covered mockup, the size of a Volkswagen, and place it in the truck's cargo bed. In spite of my disguise, a sharp security guard recognized my face, causing a real ruckus—intensified because I could not prove that I was a member of any labor union. This was very important since the company's labor contracts stipulated that only union members could physically move any equipment. A big bunch of lying ensued, the guard was sent away by his boss's boss, and finally an associate and I started for Burbank. Because of my inexperience with the multiple downshifting and double clutching needed to scale the ridge en route north, I almost destroyed the truck's transmission. All ended well, however, and our mockup fit perfectly. Since someone else returned the rented truck, I never heard (and didn't ask!) what happened to that big vehicle with its likely ground-up gears.

The Lockheed *YF-12*, with its Hughes weapon system and air-intercept missile was proven successful in only two years. The adaptation proved technically superb: in test launches, six of seven *Super Falcons* made target kills. One spectacular world record was a 1962 three-hour-long *YF-12* round trip from California to Florida. On its homeward leg, while passing over New Mexico's White Sands Proving Ground, a *Super Falcon* was launched from 74,000 feet altitude, forty-nine miles from a QB-47 Stratojet drone skimming at 500 feet above the terrain. The missile scored a direct hit! This was a fabulous first demonstration of look-down-shoot-down interception capability that had been long sought for decades.

Regrettably, like its *Rapier* predecessor, this fantastic aerospace wonder machine was cancelled in 1965 because its forecasted production and operating costs were judged to be too steep for future budgeting. Also, the government's strategic defense priority had appropriately shifted from bombers to ICBMs as the principal threat. But the good news for Hughes and the United States is that the advanced electronics concepts were carried forward and greatly embellished for the USN F-14 *Tomcat*, the USAF F-15 *Eagle*, and the USN F/A-18 *Hornet*. This golden professional career boost for me stemmed from the black nature of this project.

The personal benefits from this three-year effort were the opportunity of witnessing all parts of a complex development; learning to exercise restraint in wielding full corporate authority; having freedom to choose staff members; negotiating with the renowned Lockheed; and establishing and documenting optimal procedures for Program Management and System Engineering. The latter were for use in tackling complex new projects. These steps for planning and executing were matched to the Hughes team skills, styles, and attitudes. The new methods were little hampered by habitual past practices and stodgy government methods, as described in chapter 3. These were used in internal educational seminars and later became an industry standard.

Guiding *Tomcat* Weaponry

Great leaps forward occurred to me with dedicated active participation in the exciting evolution from *Blackbird* to *Tomcat*. It began with that Skunk Works bird being cancelled in 1965. This was followed by a DoD attempt to force the USAF and USN to cooperate by contracting for two sister versions of a single fighter-bomber baseline. This would be the first "swing-wing" fighter: leading edge forward for high combat maneuverability, and angled far back to reduce drag to enable supersonic speeds. It was called the F-111 Program, managed by the USAF. Its version was to be the F-111A *Aardvark*, and its sister being the F-111B. The USN was always uncomfortable with this arrangement since its primary performance needs were of secondary priority. As the development slowly progressed at General Dynamics Corporation (GD) in Fort Worth, Texas, the USN bird became known as the *Flying Edsel*. This was a pejorative moniker comparing it to a very unpopular Ford Motor Company automobile model. Because of our *YF-12* achievements, we won the competition for a unique weapon control system and a sophisticated new guided missile. Our contract was directly from the USN rather than a subcontract from GD.

In 1964, I became manager of the **mechanical portion of systems engineering**. Similarly to the *YF-12* project, my team set the physical design standards (this time we negotiated with the performers rather than mandating compliance), and for arranging the installation provisions with the folks in Texas. We had a strong position in negotiating the latter, since we had a direct contract with the government instead of being a supplier subordinate to the F-111 designer.

As the aircraft development proceeded, the USN became even more disenchanted because of the vehicle's overweight, poor maneuverability and acceleration, and development schedule delays. Our portion progressed well, but even though we were meeting our contracted weight objective, our equipment did place an uncomfortable burden on the fighter: over a ton of electronics, and 1,000 pounds for each of six missiles. In one dramatic meeting of USAF, USN, GD, and Hughes, I was asked about our system's capability in surface strike missions. Before I could reply, the commanding general bellowed: "not one pound for air-to-ground!" The USN eventually was allowed by the federal administration to cancel the F-111B in 1968. This triumph was dramatically done by Vice Admiral Tom Connolly, Deputy Chief of Naval Operations for Air, in a statement to Congress:

> There isn't enough power in Christendom to make
> that airplane what we want!

This enabled the USN to begin development of the F-14 *Tomcat* (that name had been Admiral Connolly's fighter call sign), to be developed by Grumman Aircraft Engineering Corporation in Bethpage, Long Island, New York. Then followed a competition for the weapon control system, with our principal adversary being the USN favorite for radars,

Westinghouse Electric Corporation. We won, based on our successes in the F-111B: we offered a system fully integrated with its primary missile. Westinghouse was hampered to create its integrated weapon development by not being in the missile business. These made our new development of lower risk and likely lower cost. That was most fortunate since the project became the largest business sector in Hughes for several years and led to our 1978 victory in the F/A-18 *Hornet* radar competition.

Tomcat Weapon System and Primary Missile
(courtesy of UNLV)

Tomcat's mission, called FAD (Fleet Air Defense), was to be the perimeter defense surrounding a Naval Task Force. Three aircraft flew elliptical patterns 100 miles from the ship formation's center. The Hughes system was to detect hostile attacking aircraft or cruise missiles another 200 miles distant, simultaneously track as many as twenty-four, recommend the highest threats, and be able to rapidly launch as many as six intercept missiles at separate targets. These could all be in flight at the same time, with those hostiles spread 50 miles apart and 60 miles away. To handle that workload, *Tomcat* had a pilot and a weapon system operator. In addition to FAD, it featured excellent dogfight capabilities for close-in skirmishes. The matured design for all those features was thoroughly proven in flight tests.

The first deployment was in 1974, with two squadrons aboard the USS *Enterprise*. In all, 712 aircraft were manufactured, including 70 delivered to Iran. Hughes delivered 6,200 *Phoenix*, with 600 going to Iran. This venerated machine, including our AWG-9 weapon control system and its matching AIM-54 *Phoenix* missile, was honorably retired in 2006, after thirty-two years of service. For those interested in thrillers, *Tomcat* played the key role in the 1988 film *Top Gun*, starring Tom Cruise as the fighter-jock hero.

Following many years on the *YF-12*, the F-111B, and the F-14 *Tomcat* startup, I was appointed Division Manager to directing the entire Hughes effort on the F-14 Weapon System. Three thousand employees were involved, as well as many subcontractors and suppliers. This rendered to me a comprehensive overview of all facets involved in complex technology, military needs and attitudes, budgeting and scheduling, and high-pressure deadline responses. Other personal gains included establishing a lasting bond with the Air Navy; perfecting how to deal amicably with Grumman; besting Raytheon in their competitive gambits; and exposure to overseas markets in Iran, Israel, and Australia. Of course, it was a thrill to observe proofs of our design: radar detection of small targets in excess of 200 miles; simultaneously tracking twenty-four; seeing the longest-range missile intercept at 120 miles launch range; and observing six missiles in flight against six targets spread by 50 miles.

My current feeling of things I contributed in this era are assuring customer satisfaction; inspiring performance excellence by our staff; demonstrating that senior managers can correctly perform rigorous flight testing; succeeding in critical Congressional hearings; and persuading Iran to purchase those six hundred *Phoenix* missiles.

Capturing and Coaching *Hornet* Radar

In 1973, the USN began a search for a new aircraft to replace the F-4 *Phantom*, A-4 *Skyhawk*, and A-7 *Corsair*, as well as to complement the heavier and more expensive F-14 *Tomcat*. Eventually evolved was the F/A-18 *Hornet*, capable of meeting multiple mission assignments. Incredibly, this diversity in Naval operations included: fighter escort, fleet air defense, air interdiction, close air support, defense suppression, and reconnaissance. McDonald Douglas Aircraft Corporation of St. Louis, Missouri, became the prime contractor, with an equal sharing of the program with Northrop Aircraft Corporation of Hawthorne, California. This was because the foundation of the vehicle design was based on the latter's YF-17 *Cobra* design that had lost to General Dynamics in the USAF fly-off to select the F-16 *Fighting Falcon* configuration. The first test flight of the *Hornet* occurred in 1978, and it was initially deployed by the USN in 1983. The combat *Hornet* has a single pilot, two engines, can fly at supersonic speeds, and has the ruggedness needed for repeated aircraft carrier operations. So far, 1,500 have been manufactured and are in use by the USN and USMC, Australia, Canada, Finland, Kuwait,

Malaysia, Switzerland, and Spain. Many improved versions have been made, and this marvelous aircraft is still very active, accumulating thirty-four years of service. It also is the hot-shot fighter used by the Blue Angels team in exciting worldwide air shows.

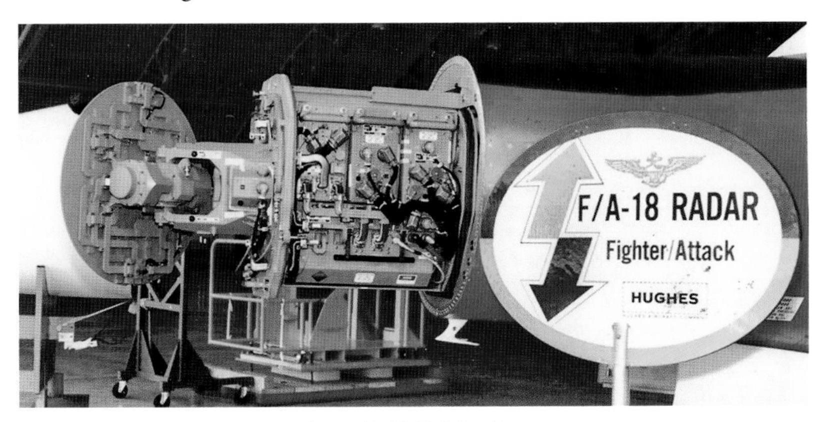

Hornet's APG-65 Radar
(personal files)

I was selected in 1976 to become **Program Manager** for our competition to capture the *APG-65* radar slated for the F/A-18. This would fill a vital Hughes business development and production need. My new title might appear as a step-down from that lofty Division Manager position. The appointment was sweetened by my also being named as one of two **Vice Presidents** of the 14,000 personnel Radar Systems Group headquartered nearby the Los Angeles International Airport. (This gain in stature was a bit offset by my incurring a miserable daily commute of one hour each way). The radar was to be subcontracted by McDonnell Douglas, and the proposal scramble was very fierce. We again faced into that very competent Westinghouse. We had just lost to them the radar for the USAF F-16 *Fighting Falcon*. Unfathomable were uncertainties about possible leanings of leaders in St. Louis and in the USN: if we won, we would carry responsibility for three of the four fighter-radar developments in the United States. Westinghouse had performed quite well for both these customers during the extensive F-4 *Phantom* Program; and we bore an image of being inherently more expensive.

Favoring us were our lead in pulse-Doppler design; greater advances in digital architecture and hardware; excellent performance demonstrated by our *Tomcat* system; and a superb bond with McDonnell Douglas for our F-15 *Eagle* radar work. The proposed new design would feature the first programmable digital signal processor, flexible multimode capabilities, high gain antenna, and the ability to track targets using any approach angle. The pleasures for me in this assignment were winning the difficult competition; learning to interact properly with the aircraft prime as our customer rather than the government; at the same time maintaining our strong USN bond; and working directly with the world's leading radar engineers. This program became the largest production business in the history of Hughes Aircraft.

One grievous negative at this time was observing the single case of espionage the company experienced during its seventy-year existence. The perpetrator passed to the Russians the complete initial proposal, including lengthy descriptions of our unique design. (The first page of each volume contained a letter signed by me, so perhaps a personal focus was placed on me by the KGB!) Fortunately, Bill Bell was caught and convicted, but we had to do massive redesign to keep the radar still viable facing the U.S.S.R. countermeasures. Nothing could be done to retrieve the extensive technology transferred that had been perpetrated.

Here are my perceptions of what I contributed on this project: standing my integrity ground by refusing during the competition to sign up to specifications that could not be met; achieving a most effective and efficient program management and systems engineering operation; meeting all schedules; educating the team not to freebee performance upgrades without contract price adjustment; greatly helping the successful sale to Spain; and personally inventing antenna vertical scan for dogfight effectiveness (now incorporated in every combat fighter).

Honchoing Guided Missiles

The biggest demand on my skills and the premier learning opportunity was being selected in 1983 by CEO Dr. Allen Puckett to become **President of the Missile Systems Group** (MSG). At that time, I was still one of the two VPs of the Radar Systems Group. The big mountain for me to climb was to be effective as a responsible leader of over 13,000 personnel, including 3,500 engineers and scientists in every field of technology. The MSG contributed a strong portion of the overall company's sales and earnings, nurtured technology advancement in several vital fields, and bore the heavy burden of keeping our prestigious world leadership in intelligent weapons. The products we produced were essential elements of the U.S. defense base, and contributed much to settling the Cold War against the U.S.S.R. without combat. We also supplied guided weapons to fifty other friendly nations.

At the time of my appointment there were two principal operating sites in the West, with recent manufacturing expansion into Georgia and

Alabama. A team of about 3,800 was based in Canoga Park at the west end of California's San Fernando Valley. Most of the design, development, and test verification of our products was performed there; it also housed the executive leadership offices.

At Canoga Park, about twenty advanced systems were under development, some protected by top-secret security. The MSG needed expertise in virtually every sector of science and engineering.

Author with CEO Dr. Allen Puckett
(personal files)

Here are some clues to gain perspective on the complexity of these products. Missiles can include an airframe, a rocket motor, wings, steering vanes, control mechanisms, computers, electrical and hydraulic power supplies, cooling systems, warheads, proximity fuses, and intelligent seeker heads. Establishing a design configuration demands much careful thought and analysis. Nothing devised to date can solve every encounter problem that may arise to meet a targeting objective. A plethora of variables include the likely combat situation, weather, ease of use, target vulnerability, possible countermeasures, collateral damage avoidance, and carrier vehicle constraints. All these are piled atop concerns for the costs of development, manufacturing, and lifetime support. Successful guidance methodologies include: commands from the launcher or a satellite; tracking of reflections made by active illumination; homing on electromagnetic or sonic waves emanating from an object; and inertial sensing and memory updated by GPS (Global Positioning System).

Our Tucson manufacturing base was opened in 1951. The geographical location had been endorsed by the U.S. Defense Department, which at the time wanted most military hardware production plants to be remote from any ocean coastline, thereby minimizing vulnerability to possible hostile air attack. In the 1980s, its staff numbered 8,500, and it was ranked as the largest private employer in Arizona. Products being manufactured in 1983 were anti-tank *TOW*; air-to-ground *Maverick*; air-to-air *Phoenix*; surface-to-air *Roland*; glide bomb guidance heads for *GBU-15* and *Walleye*; bomb deliver cueing; and launcher mounts for these weapons.

One giant mountain for me to climb was to gain acceptance by the thousands of MSG employees, beginning with surviving my first meeting with thirty senior managers in Canoga Park. I already knew them all personally because of my long involvement with the USN *Phoenix* missile. One very able but crusty old-timer boldly stated, "We don't need some slick operator from downtown coming here to tell us what to do. We're doing just fine already." After a moment of shocked silence, all I could think to say was, "Well, I'll give it my best shot." My usual behavior had been to quickly forgive and forget, but that public slap in the face took a long time to heal. Although we all frequently and openly spoke our private opinions (we enjoyed being mavericks), this comment was great contrast to the traditional Hughes family relationship. The subsequent five years were very stressful but did not seem to include any rejection by key members of this remarkable team of diverse performance winners.

Acceptance by the manufacturing folks at Tucson was even more difficult. That organization felt it should be reporting directly to the corporate CEO, as it had in 1951, rather than to an intermediate executive gang in distant Canoga Park. The outfit was nicknamed "The Cactus Curtain," due to its unwillingness to reveal inner operating difficulties to Group leaders. That separatist attitude changed radically when Tucson, Canoga Park, and corporate Culver City were intimately teamed for survival during the manufacturing quality turmoil described in chapter 5.

During my five-year tenure, there were many major difficulties in keeping MSG on course to fulfill its obligations to corporate as well as to its many customers. Some came as a surprise, some because of changing external situations, others from long-standing poor internal operating procedures. Being MSG President was indeed the most rewarding segment of my life at Hughes. In retrospect, as many retired folk may feel about their job performance, I wish I had performed big responsibilities with more spunk and "true grit." The results might have been more stellar. However, I do feel that during this time I helped our team in these ways: inspire everyone to excel at bettering our goals; practicing and urging comprehensive integrity; success in besting our competitors, strategizing recovery from quality turmoil, and achieving long-term customer satisfaction.

SUMMARY

CITIZENS OF THE U.S. should be very concerned about our nation falling behind many other countries as sources of advancement in virtually all technologies. During the last century, we prevailed in first place in almost every field of innovation. This was a principal reason for our expansive and robust economy. We were aided by exploiting vast amounts of natural resources. But now the mighty competitive struggles are over the uses and improvements in knowledge. We now rank no higher than 4th in creativity, 41st in education quality and content, and an abysmal 85th in the proportion of graduates in science and technology (in the latter category we are at 5 percent, the EU at 15 percent, and China at 21 percent). Unfortunately, many of the technical graduates in U.S. universities are from other countries; they return home to apply their skills to benefit our global competitors.

Reversals of these negative trends can be achieved by intensive actions in government and industry, coupled with a supportive and eager public. The government should implement these positive actions:

1. Persuading the population of the urgency to place priority on science and engineering
2. Increasing resources devoted to education
3. Enhancing the achievement standards in our school systems
4. Substantially multiplying investment in research and development

It is noteworthy to point out that a recent Boston Consulting Group analysis revealed China is investing twice as much per year on research and development as the U.S.; their 2018 budget for that is $658 billion.

It is also alarming that their focus is on development rather than research, often using inventions stemming from U.S. investments. This gives them a competitive edge by more quickly bringing new commercial products to market. A measure of this also is that U.S. manufacturing plants now are operating at less than 80 percent of capacity, compared to over 90 percent a decade earlier. Of course, this output shift to Asia is the source of a massive revenue accumulation in foreign coffers.

The most productive way to apply government R&D funding is by establishing sets of national goals, defining projects directed toward those objectives, and competing them among industrial firms. The drive for competitive survival breeds enthusiastic staff motivation, increases productivity, and fosters free-thinking teamwork searching for technology breakouts.

Many people feel that everything worthwhile has been invented, so allocating budgets to R&D would be wasteful. They apparently are unaware of our nation falling behind others. A profound statement in 2002 by Donald Rumsfeld could enlighten the doubters:

> There are known knowns. There are things that we know we know. There are known unknowns. That is to say, there are things that we know we do not know. But there are also unknown unknowns. There are things we don't know we don't know.

That means that there will always be new exploratory probes in countless directions.

Industry can do its part by enthusiastically responding to those competitions, greatly expanding corporate R&D budgeting, setting long-term goals, and attracting and rewarding technical talent.

The public can assist by supporting this new priority thrust, urging local schools to implant and sustain STEM (Science, Technology, Engineering, and Mathematics) programs, and encouraging their children to participate in practical science as a lifelong profession.

And individuals can also help accelerate this important new national mission; attain the best education in science and engineering; seek professional careers directed toward advancing technology; increase interest in these goals among friends and associates; and help stimulate public recognition and rewards to those successfully demonstrating creativity.

This book may also help to properly implement and expand that new national push to regain our place in this competitive world.

We have examined in these chapters a multitude of innovations achieved in past eras. It is appropriate to consider these wonders not only through learning how they function and help our lifestyle, but also to attempt to fathom the likely thinking processes of those inventors: what motivated them, what inspired them to break through perceived boundaries and persist until the device became practical. If we can understand those ingenious thinking processes, we may encourage others to mimic them in order to jump to new heights in creativity.

Entrepreneurs motivated to establish new organizations and teams to perfect new products can learn from this text of the necessary steps to be taken. It is fruitful to set up a workplace conducive to assuring comfort, stability, and support for the workforce. When such an ambience has been secured, the leaders must establish vital pillars of operation: a philosophy that nurtures free thinking, accepts risks, adjusts to short-term errors, and amply recognizes and rewards individual contributions. They must also embody a meaningful long-term mission and inspire the teams to progress toward those goals.

To assure long-term viability, the enterprise must gird itself for continual competitive challenges, respond and adjust to customers' whims and desires, assure sufficient revenue for ongoing financial stability, and employ fundamental integrity in business and technical activities.

Attracting, selecting, sustaining, and upgrading talented individuals to make up corporate teams involves continuing intensive focus. The most productive types usually have a variety of technology understandings, an innate curiosity, are self-driven to practice free-thinking, and are eager to outperform others. At the same time, they will blend with team members and share their thoughts and expertise. Since technology knowledge of all types is rapidly changing, the corporate leaders must assure adequate educational opportunities for all staff members.

A meaningful key to success is fostering open minds not encumbered by rigorous procedures. Hughes did this well, aptly describing its staff as "a large bunch of anarchists bonded only by a common parking lot."

Many examples of inventions emanating from surprising sources illustrate that everyone should listen to and observe ideas from anywhere, not just from staff professionals. An illustration not previously described is the emergence of the first telescope. German-Dutch eyeglass maker Hans Lippershey in 1608 observed his sons playing with two surplus lenses. They all noticed that when the disks were aligned and spaced apart objects in the field of view were significantly enlarged. The discovery was not his concept or results of his experiments; it was the children's random frolicking that created a new wondrous device.

Organizations matching these characteristics will surely help enable our nation to recover its leadership status in the ever-changing world.

ACKNOWLEDGMENTS

THE AID AND support of many excellent people enabled this book to mature. Their efforts are greatly appreciated; without them the text could not have become possible.

- PUBLICATION
  Greg Sharp, President, Sea Hill Press

- PROFESSIONAL EDITING AND LAYOUT
  Cynthia Sharp, Sea Hill Press

- ADVISING, TRANSCRIBING, AND EDITING
  Charlotte Richardson

- CRITIQUE, EMBELLISHMENT, CONTENT PROOFING
  Dr. Art Chester, former Senior Vice President and Director of Research Laboratories, Hughes Aircraft Company
  Ed Cobleigh, former Missile Systems Marketing Manager, Hughes Aircraft Company
  Greg Hartley, Electrical Engineering Department Manager, Raytheon Vision Systems
  Bruce Richardson, President, Queo Systems Ltd.

- SEGMENT EDITORS
  Dick Godfrey, former Managing Director, Trust Company of the West
  Lou Kurkjian, former President, Ground Systems Group, Hughes Aircraft Company

Milt Radant, former Senior Vice President for Technology,
Hughes Aircraft Company
Bob Sendall, former Senior Scientist,
Hughes Aircraft Company

- Author of Foreword

Dr. Yannis Yortsos, Dean of the USC Viterbi School of
Engineering

- Names of Those Lending Recent Quotes

Vice Admiral Robert Baldwin (USN), former Commander
U.S. Seventh Fleet; Deputy CNO for Air
Dr. John Cashen, former Vice President,
Northrop Grumman
Dr. Art Chester, former President, HRL Laboratories;
former CTO for Hughes Electronics
Ed Cobleigh, former Director of Marketing,
Hughes Aircraft Company
Vice Admiral Thomas Connolly (USN), former Deputy
CNO for Air
Lieutenant General Walter Dürig, former Commander,
Swiss Air Force
Lieutenant General Donald Lionetti (USA), former Director,
Ballistic Missile Defense
General Richard Myers (USAF), former Chairman,
Joint Chiefs of Staff
Dr. Allen Puckett, former CEO, Hughes Aircraft Company
General H. Norman Schwarzkopf Jr. (USA), former
Commander, Desert Storm
Ron Shaffer, former Manager, Global Marketing,
General Motors
Gloria Wilson, former Senior Engineer,
Hughes Aircraft Company

- Fundamental Foundation of Professionalism

An infinity of thanks go to the Hughes Aircraft Company family team for teaching me the wonders of electronics. Hundreds of creative professionals, advisors, mentors, inspiring leaders, geniuses, hands-on workers, and supporting staffs provided a global view of proper organizational behavior and practices.

REFERENCES

Arthur, Brian W. *The Nature of Technology: What It Is and How It Evolves.* New York: Free Press, 2009.

Blanchard, Kenneth H., and Spencer Johnson. *The One-minute Manager.* New York: Morrow, 1982.

Carnegie, Dale. *How To Win Friends and Influence People.* New York: Simon & Schuster, 1936.

Cornell University, INSEAD, and WIPO (2016): *The Global Innovation Index 2016: Winning with Global Innovation.* Ithaca, Fontainebleau, and Geneva, 2016.

Desai Meghnad. S. Fukuda-Parr, C. Johansson, and F. Sagasti. *Measuring Technology Achievement of Nations and the Capacity to Participate in the Network Age.* Journal of Human Development Vol. 3, Iss. 1, 2002.

Dictionary.com, *Dictionary.com.*, 2017.

Diebold, John. *The Innovators: The Discoveries, Inventions, and Breakthroughs of Our Time.* Dutton, 1990.

Drucker, Peter F. *The Essential Drucker Selections from the Management Works.* London: Routledge, 2011.

Dweck, Carol S. *Mindset: The New Psychology of Success.* New York: Random House, 2006.

Economist Group. *The Economist Magazine.* "Gender Pay Gap." Dorrance Publishing, October 7, 2017. https://www.economist.com/news/international/21729993-women-still-earn-lot-less-men-despite-decades-equal-pay-laws-why-gender.

Encyclopedia Britannica, *Internet Britannica.com,* 2017.

Isaacson, Walter. *The Innovators: How a Group of Hackers, Geniuses, and Geeks Created the Digital Revolution.* New York: Simon & Schuster, 2015.

Jahren, Hope. *Lab Girl.* New York, Alfred Knopf, 2016.

Johnson, Steven. *How We Got to Now: Six Innovations that Made the Modern World.* New York: Riverhead, 2014.

Korn Ferry Hay Group. "The real gap: Fixing the gender pay divide." New York: Korn Ferry, 2016. https://www.aesc.org/sites/default/files/uploads/documents-2015/KFHGGenderPayGapMay2016.pdf.

Mukherjee, Siddhartha. *The Gene: An Intimate History.* London: The Bodley Head, 2016.

National Geographic Magazine. June 2017 Issue. Washington, D.C.: National Geographic Society.

Pande, Peter S., Robert P. Neuman, and Roland R. Cavanagh. *The Six Sigma Way: How GE, Motorola, and Other Top Companies are Honing Their Performance.* New York: McGraw-Hill, 2001.

Pearl, Matthew. *The Technologists.* New York: Random House, 2012.

Richardson, D. Kenneth. *Hughes After Howard: The Story of Hughes Aircraft Company.* Santa Barbara: Sea Hill Press, 2011.

Rose, D. and DJ Pevalin. "The National Statistics Socio-economics Classification Unifying Official and Sociological Approaches to the Conceptualisation and Measurement of Social Class." ISER Working Paper no. 2001-4. Colchester: University of Essex, 2001. https://pdfs.semanticscholar.org/43b4/3711d6c1d2cc7cbe691cd5e935e2122991fa.pdf

Sirkin, Hal, J. Rose, and R. Choraria. *An Innovation-led Boost for US Manufacturing.* The Boston Consulting Group. Apr 17, 2017. https://www.bcg.com/publications/2017/lean-innovation-led-boost-us-manufacturing.aspx.

Sloan Jr., Alfred P. *My Years with General Motors.* New York: Doubleday, 1963.

Tredgold, Gordon. *Leadership: It's a Marathon Not a Sprint–Everything You Need to Know about Sustainable Achievements.* UK: Panoma Press, 2013.

BREAKTHROUGHS BY HUGHES AIRCRAFT COMPANY

Premier High-Performance Devices and Inventions in Electronics

Some readers may be interested in scanning this list of unique creations by Hughes. This compilation was assembled by the author in 2016. Further details may be found in *Hughes After Howard*, published by Sea Hill Press in 2011.

As noted in earlier chapters, nearly every major advance in the history of science and technology has relied on the creativity and hard work of many inventors and engineers over a large time span. Although an issued patent may list only one or a few leaders, the useful application of the invention typically depends on development by many others.

It is in that spirit that these inventions and developments have been listed. In some cases, Hughes had the idea, in other instances another company or source may have had the first notion. However, in each instance Hughes people made major contributions to the refinement, maturation, and practical application of the technology. The breadth of contributions listed here illustrates the immense variety of areas to which a large technical organization can contribute.

- RADAR
  Clutter avoidance using pulse-Doppler
  Digital signal processing
  Programmable signal processing
  Multiple target tracking

Interleaved pulse repletion frequency types
Planar disk antenna
Artificial antenna array for high resolution ground mapping
Electronic scan and beam steering
Active array modules
Superior performance in stealth

- INFRARED
  Mercury-Cadmium-Telluride sensitive & broadband sensors
  Serial scan imaging
  Multi-matrix image scanning
  Unique sensor cooling devices
  Motion-stabilized IR and laser tank gunsight
  Long distance ICBM search and track

- SONAR
  High resolution towed arrays
  Long-range high resolution object imaging
  Extreme sensitivity passive tracking and ID
  Imaging target tracker for torpedoes

- LASERS & OPTICS
  First operational laser
  Q-switch on-off control
  Precise range measurement
  Secure communication
  Directed energy weapons
  Precise stabilized, adaptable IR, laser, and visible target tracking

- GUIDED MISSILES
  Proportional navigation
  Tail fin steering
  Look-down-shoot-down using pulse-Doppler guidance
  Time-shared semi-active radar guidance vs multiple targets
  Active internal radar
  Wire command guidance for anti-tank and torpedoes
  Longest range air-to-air intercept
  Air-to-ground guidance using internal infrared, laser or TV imaging

Star tracker guidance
Global-range inertial guidance
High altitude ICBM intercept

- Computer Networks
  Air defense command-and-control for large nations and NATO
  High performance Air Traffic Control
  Real time location of combat units in large areas
  Ballistic missile launch detection, viewing areas the size of Kansas
  Projectile tracking and precise artillery counter-fire
  Integration and analysis of multiple imaging devices

- Spacecraft & Satellites
  Fully instrumented soft landing on moon
  First TV images of Earth
  Venus atmospheric probes
  Multispectral environmental earth images
  Ion engine deep space thrusters
  Geo-synchronous orbiters
  Spin-stabilized satellites
  De-spun core preventing tumbling of large satellites
  Aspect control by Sun tracker
  High bandwidth relay responder modules

- Microelectronics
  Thin-film wafer fabrication
  Submicron lithography
  Self-aligned gate arrays
  Silicon-sapphire semiconductors
  Ion implantation
  Gallium arsenide amplifiers

- Derivatives That Improve Human Lifestyle
  Global communication by satellite
  Television broadcast by satellite (*DirecTV*)
  OnStar wireless automotive services (security, diagnostics, navigation, communication)
  Cell phone processor architecture

Lasers in medicine, manufacturing, science, communication, and entertainment
Digital watches
Night vision
Single wire diverse data stream transmission
Clutter-free radar
Micro-electronic device shrinkage and performance increases
3-dimensional sound from only two speakers

D. Kenneth Richardson, author of
Hughes After Howard and *Sparking Innovation*

ABOUT THE AUTHOR

KEN RICHARDSON, AUTHOR of *Sparking Innovation* (Sea Hill Press, 2018) and *Hughes After Howard: The Story of Hughes Aircraft Company* (Sea Hill Press, 2011), was raised in Hawaii and began his career at Hughes Aircraft Company as a radar design engineer.

In forty years of service, Ken rose to become president of Hughes' Missile Systems Group and then corporate president and chief operating officer. This team of more than 80,000 employees created products at the forefront in every field of electronics technologies.

Ken received degrees with honors in engineering and business administration from Tufts University, USC, and UCLA. In the early 1990s, each of these institutions, as well as the Los Angeles Unified School District, selected him for its annual award for leadership excellence. He has served as trustee on the boards of six large nonprofit organizations.

An active life has included raising two sons, sailing, scuba diving, golfing, backpacking, earning a private pilot license, hands-on flights in many military aircraft, and visits to over one hundred nations.

After retiring in 1991, he now resides with his wife, Charlotte, in Santa Barbara, California, and enjoys activities with four grandchildren.

Hughes After Howard: The Story of Hughes Aircraft Company
By D. Kenneth Richardson

ALSO BY KEN RICHARDSON: *HUGHES AFTER HOWARD*

PEOPLE EVERYWHERE HAVE heard of the eccentric Howard Hughes, but few know that in 1953 he virtually disappeared from the company he had begun in 1932. Under new creative and inspired management, Hughes Aircraft Company became the leading military electronics organization in the world and rose to 85,000 employees. Some called it a national treasure.

In *Hughes After Howard: The Story of Hughes Aircraft Company* the company's past president Ken Richardson shows how this was done. Collaborating with over ninety retired employees, Ken has compiled this remarkable piece of American aerospace history. Learn about many complex products in all fields of electronics crafted by this highly motivated, inventive team.

> *"Howard Hughes and his brainchild, the Hughes Aircraft Company, laid the technological foundation for American success in prevailing in and ending the Cold War."*
> —Tom Reed, former Secretary of the Air Force.

Hughes After Howard: The Story of Hughes Aircraft Company (Santa Barbara: Sea Hill Press, 2011) by D. Kenneth Richardson can be purchased through seahillpress.com